THIS
ECSTATIC
NATION

THIS
ECSTATIC
NATION

The American Landscape

and the

Aesthetics of Patriotism

TERRE RYAN

University of Massachusetts Press

AMHERST AND BOSTON

LC 2011021557
ISBN 978-1-55849-873-0 (paper); 872-3 (library cloth)

Designed by Sally Nichols
Set in Janson Text
Printed and bound by Thomson-Shore, Inc.

Library of Congress Cataloging-in-Publication Data

Ryan, Terre, 1959–
This ecstatic nation : the American landscape and the aesthetics of
patriotism / Terre Ryan.
p. cm.
Includes bibliographical references and index.
ISBN 978-1-55849-873-0 (pbk. : alk. paper) — ISBN 978-1-55849-872-3 (library
cloth : alk. paper) 1. West (U.S.)—Environmental conditions. 2. Landscapes—
Symbolic aspects—West (U.S.) 3. Landscapes—Social aspects—West (U.S.) 4.
Nature—Effect of human beings on—West (U.S.) 5. Land use—Environmental
aspects—West (U.S.) 6. Landscapes in art. 7. West (U.S.)—In art. 8. Nuclear
weapons—Testing—Environmental aspects—Nevada—Nevada Test Site.
9. Clearcutting—Environmental aspects—Oregon. 10. Ranching—
Environmental aspects—Wyoming. I. Title.
GE155.W47R93 2011
363.700978—dc23
2011021557

British Library Cataloguing in Publication data are available.

*For Elizabeth, Christina, and Jack
and in memory of Janet Marek-Ruf*

This ecstatic Nation
Seek – it is Yourself.
EMILY DICKINSON

CONTENTS

ILLUSTRATIONS

ACKNOWLEDGMENTS

This book was long in the making, and I am grateful for the help I received from colleagues, friends, and family. At the University of Nevada, Reno, I extend warm thanks to Scott Slovic and Cheryll Glotfelty for their encouragement and support at early stages. My thanks as well to Scott Casper, Ann Keniston, and Nancy Markee for their helpful input. I am extremely grateful to Ann Ronald for her support through the James Q. and Cleo K. Ronald Memorial Dissertation Fellowship.

Other teachers and mentors who made a difference on my long and circuitous journey include Debra Earling, Philip Eliasoph, Dan Flores, Brady Harrison, Wayne Koestenbaum, the late William Matthews, Kitty McDonough, Leo F. O'Connor, and Barry Wallenstein. I carry you with me.

Kathleen Dean Moore steered me toward the Oregon coast; Don Snow pointed me toward Pinedale, Wyoming; and SueEllen Campbell provided important early directions. I thank them for navigating.

I thank the staff of the U.S. Forest Service's H. J. Andrews Experimental Forest for giving me the opportunity to conduct field research. Kathy Keable, Al Levno, Kari O'Connell, Fred Swanson, and Debbie Wiseman enabled and enriched my experience at the Andrews. I am especially grateful to Fred Swanson for his help.

Special thanks go to Lee and Gennie Barhaugh and the crew at the Nature Conservancy's Pine Butte Guest Ranch, who welcomed a tenderfoot and showed me angles on Montana life and landscape I wouldn't otherwise have seen. I thank Judy Singer for providing friendship and shelter.

I thank Erin McKeen at the U.S. Department of the Interior and Sara Frantz of the Nevada Museum of Art for their efforts in supplying some of the illustrations in this text. Charlotte Labbe and the reference librarians at Fordham University and the University of Nevada, Reno, provided invaluable assistance. Chester Arnold gave generously of his time.

My friends helped in various ways. I extend heartfelt thanks to Mark and Amy Bennett, Rene and Greg Blair, Michael Branch, Miriam Burns, Hope Freeman, Dave Johnson, Jan Mowder Hughes, Donna and John McNeil, Theresa Leininger-Miller, Colleen O'Brien, Sarah Perrault, Suzanne Roberts, Chris Robertson, Rebecca Rosenberg, Janet Marek-Ruf and Ed and Skyler Ruf, Tracy Sangster, Michelle and Tony Seymour, Tammie and Buddy Smith, Tamara Sevransky, and Jennifer Hughes Westerman. I am eternally grateful to superhero Paul Knox, who did the math. Touchstones Sue Roughley and Bill and Jeri Russell offered insightful comments on early versions of the manuscript.

I am extremely fortunate to have had the opportunity to work with the staff at the University of Massachusetts Press. I extend my sincere thanks to my editor, Brian Halley, for opening the door. Special thanks go to Carol Betsch and Mary Bellino for their insightful comments and their help in crafting the final version of this manuscript. I thank Sally Nichols for managing the design. I am also very grateful to the anonymous readers for the press; their comments on earlier versions of the manuscript enriched this work.

For their support, I thank my brothers and sisters, Mary Lou, Francisca, Juan, Patsy, and Michael, my brother-in-law, Jim, my nieces, Elizabeth and Christina, and my nephew, Jack. Love endures time and distance.

The epigraph for this book is taken from Emily Dickinson's poem F1381, "The Heart is the Capital of the Mind," reprinted by permission of the publishers and the Trustees of Amherst College from THE POEMS OF EMILY DICKINSON: VARIORUM EDITION, edited by Ralph W. Franklin, Cambridge, Mass.: The Belknap Press of Harvard University Press, Copyright © 1998 by the President and Fellows of Harvard College. Copyright © 1951, 1955, 1979, 1983 by the President and Fellows of Harvard College.

Portions of Chapter 1 appeared in "Creation Stories: Myth, Oil, and the Arctic National Wildlife Refuge," *Journal of Ecocriticism* 2 (January 2010): 81–86.

THIS
ECSTATIC
NATION

Introduction

~~≈~~

We need a new story.
William Kittredge, *Who Owns the West?*

In December of 1997, desperate to live out West, I traveled from my home in New York City to Washington, D.C. A casual observer might have remarked that I needed a better map, but I was headed for the National Gallery of Art and the first retrospective of the work of nineteenth-century landscape painter Thomas Moran. Moran was only one of the luminaries in an era when Americans, seduced by their country's beauty, fell in love with their landscape. American landscape imagery, wildly popular, graced the walls of galleries and homes and the pages of magazines and newspapers. Landscapes of all regions came into vogue, but dramatic western scenery particularly captivated the public, and such images became the aesthetic backdrop of expansionist mythology. Although art and taste have certainly evolved, the national love affair with landscape imagery has never ended, and that December day, this exhibition was as close as I could get to my panoramic dreams.

Many of Moran's works were crafted according to an aesthetic of awe that art historians call "the sublime." Early theorists, artists, and much of the public believed that forbidding landscapes revealed the formidable power of God. Moran is best known for his images of what would become Yellowstone National Park, but among the most striking (to me) of the works in this exhibit was his 1875 *Mountain of the Holy Cross* (shown in figure 1), which portrayed a snow-filled cross near

the peak of a Colorado mountain. The curator's note indicated that
the public generally perceived the mountain's cross as a sign of God's
blessing on the nation's expansionism. Since the Christian cross originally symbolized humility and sacrifice, and since American empire
building included slavery, land theft, genocide of Native Americans,
systematic abuse and marginalization of ethnic minorities, the deaths
of emigrants duped by the promise that rain really did follow the
plow, and massive environmental degradation, the curator's note was
troubling.

But this was still the Manifest Destiny era, when expansionism
was largely promoted as divine mission and the West was still the
frontier. The American studies scholar Richard Slotkin observes that
"Frontier Myth and its ideology are founded on the desire to avoid
recognition of the perilous consequences of capitalist development
in the New World, and they represent a displacement or deflection
of social conflict into the world of myth."[1] Many of the landscape
images that post–Civil War painters and engravers produced reflect
what I have come to call *Manifest Destiny aesthetics*, which project the
mythology of the expansionist era, romanticize the landscape, and
obscure the violent stories of those places. The cross on the mountain
elevated expansionism into mythic space suffused with nationalism
and sublime authority that expansionist-era boosters attributed to
God. While Yellowstone country has its own complicated stories of
displacement of indigenous peoples, development of resources, and
animal abuse, I found it easier to get lost in Moran's watercolor and
oil paintings of "Wonderland," which had so charmed the nation that
Congress designated Yellowstone as our first National Park in 1872.[2]

Six months later I was in Yellowstone country for real, where I
paused en route to Montana. If I had driven into my dreamscape, I
woke up in Yellowstone. I made all the right tourist's moves, like taking pictures of Old Faithful, visiting areas that Thomas Moran had
painted, and logging hours in my car. I bounced over potholes and
idled in clots of traffic so dense I decided that Yellowstone was New
York City with elk. Things got better when I drove out of the park,
toward my new home. A gentle rain tapped the windshield, the sun
was to my back, and—I'm not making this up—a rainbow straddled

the road. I was listening to a composition of guitar, flute, and storm sounds by string-musician William Eaton and Native American flutist Carlos Nakai, so the scene even had the right soundtrack. The rainbow looked like the gateway to the Promised Land, and since I wanted to believe that I was driving toward a destiny as beautiful as the landscape, I took the rainbow as a sign. And then, as if on cue, a pebble torpedoed my windshield and left a crack the size of a quarter.

The pebble provided a reality check. I was hardly the first to invest the American landscape with my own longings. My immigrant mother and paternal great-grandparents had done as much, as had the emigrants who pushed westward during the expansionist era, just as others had done since Europeans first invaded the Americas. Nor was such dream-driven migration simply archaic. The West was the United States' fastest growing region between 1990 and 2006, and Nevada, where I would eventually make my home, was the fastest growing state throughout the last four decades of the twentieth century.[3]

I saw hope in the landscape because American landscape mythology had taught me to look for it there. Thomas Moran's *Mountain of the Holy Cross* had troubled me the first time I saw it, but renditions of that image were everywhere in late twentieth- and early twenty-first-century culture, often with a sport-utility vehicle taking the place of the cross and cruising over mountains that would have struck our pioneer forebears as treacherous. Aesthetics evolve, and since the late nineteenth century, renditions of the sublime have frequently reflected what critics call the "technological sublime," which reveals the formidable power of human engineering.[4] Images like the SUV ads linked drivers to the nation's pioneer heritage, deified technology, and lifted consumerism to mythic space, where its true stories—for example, the environmental costs of building and driving the SUV—were obscured by the product's glitter and the afterglow of nineteenth-century landscape mythology.

Robert F. Kennedy Jr. observes that nature "is the infrastructure of our communities."[5] What, then, are the costs of framing the environment according to inherited aesthetic standards, and then employing these aesthetics in the service of commercial enterprises and national mythology? Scholars have documented the influence of landscape imagery on the development of American nationalism during the nine-teenth century.[6] How are landscape aesthetics informing our sense of national community and the patriotism that the historian Merle Curti, in his early studies on the subject, defined as "love of country, pride in it, and readiness to make sacrifices for what is considered its best interest"?[7] How do landscape aesthetics work with representations of patriotism, which, throughout the twentieth century and into the new millennium, have drawn upon landscape, war, and commerce?

I contend that nineteenth-century landscape aesthetics remain so influential in contemporary culture that they may encourage us to view our environment much as expansionist-era boosters viewed it—as a stockpile of resources to be exploited or as scenery to be preserved. This aesthetic framework perpetuates Manifest Destiny–era ideas of the environment as commodity, scenery, and cultural trashlands, and it romanticizes or conceals acts of environmental violence.

Given the pervasiveness of landscape mythology in contempo-rary culture, how might one learn to see critically, or ecocritically? Ecocriticism began as a branch of literary studies examining "the rela-tionship between literature and the physical environment," but we might think of it as the study of the correspondence between human beliefs and activities and the physical world.[8] As the art historian Allan C. Braddock and the cultural theorist Christoph Irmscher note, eco-criticism "emphasiz[es] the particular ways in which human creativ-ity—regardless of form (visual, verbal, aural) or time period (ancient, modern, postmodern)—unfolds within a specific environment or set of environments, whether urban, rural, or suburban." An ecocritical perspective, they add, "mak[es] us see things anew—and perchance more ethically—in their relation to the environment."[9]

Think of landscapes as the rock on which cultures inscribe their complicated stories. *This Ecstatic Nation* is an ecocritical memoir that traces my journeys around bombing grounds in Nevada, logging sites in Oregon, and energy fields in Wyoming—all scenes of extreme environmental violence on contested ground. These places serve as starting points for explorations of the overlapping trails of national mythology, landscape aesthetics, patriotic discourse, and public pol-icy. These regions have long figured in American frontier mythology and patriotic discourse, and they reveal much about how business and government agencies choose to frame our national stories.

Stories of Manifest Destiny aesthetics are not limited to these places or to the West; they also extend into virtual space, a frontier of mythic space where the *virtual sublime* depicts humans in control of technology, including technology that our engineers have yet to master. The stories of these landscapes reach beyond their regions and our borders. Weapons testing, clearcutting, and energy extraction exert dramatic regional impact, but they also contribute to the larger stories of consumption and global warming, affecting the lives of regional, national, and global communities in intimate ways. While I chose these sites, in part, because I could reach them from my home, the activities in these places—including the bombing grounds, according to the dominant cultural discourse about war—enable my lifestyle. So I examine my complicity in a system that still, like the nineteenth century's Manifest Destiny art, often links exploitation of the land to progress and patriotism. I drove to each place in a 1998 Toyota Corolla. It was the most efficient car I could afford, but the footprints and tire tracks I left in my wake are deeper than they should be. And I am complicit in environmental destruction in far more ways than those I explore here.

These journeys raised questions about my nation and my lifestyle that were sometimes difficult to face. But as the Oregon historian David Peterson del Mar remarks, "Mythic, uncritical history impoverishes and distorts both our past and our present. . . . To care about a place entails building on its strengths and weaknesses."[10] I found great cause for hope in the work of scientists, artists, historians, and other original thinkers who continue to tell us the truth about our actions and who strive to find the means by which we might step more lightly on the land we claim to love. Their work points toward what we might call *green patriotism*, an ethic of caretaking that transcends outdated divide-and-conquer landscape mythology and political party lines.

Environmental concerns are not partisan affairs, although they are often represented that way. Evolving Euro-American aesthetics have figured in the framing of the American landscape and in the conquest and use of natural resources since the first European settlers arrived here. But as the western writer William Kittredge observes of American landscape mythology, "We need a new story."[11] We also need new ways of looking at the land. I hope to increase public awareness of the costs of representing our environment through the aesthetic lens of the empire-building era.[12]

1
Pass the Bottle
Scenic America

The health of the eye seems to demand a horizon.
Ralph Waldo Emerson, *Nature*

My first awareness of landscape as a package of values imposed on a portion of earth came when I was still quite young, and since I was an American kid from the suburbs, I experienced this great awakening in the family car. I was ten or so, probably carsick, and crammed into a station wagon with my entire family for a Sunday drive. Ours was a Ford Country Squire, a name that implied an aristocratic version of the Jeffersonian pastoral. That image was entirely out of sync with the lemon-colored station wagon—furnished with 1960s state-of-the-art faux-wood side paneling and black vinyl interior—and the family that actually used it. I was stationed beside a window—the sole perk that carsickness afforded me—so that, if the unthinkable happened, I could aim at New Jersey rather than my siblings. My older sister called the other window, one sister got stuck with the hump in the middle, and the three youngest kids, in that pre-seatbelt era, were tossed into the back like a litter of puppies. I liked this arrangement because I could look out at the world as we hummed along the Garden State Parkway.

The parkway is an asphalt river that flows between southern New Jersey's sandy pines and the birches, oaks, and maples of the state's northern regions; in the mid-1960s, the federal government attributed the parkway's admirable safety record to its beautiful landscaping. From a kid's perspective, the parkway was fine until we reached the border of Newark, East Orange, and Irvington, where the highway suddenly slashed straight through a massive cemetery. There were so many tombstones that it looked as if we were driving through a dwarfed city, but it was a city whose inhabitants were all dead.

To cold war kids who expected a nuclear attack at any moment, this deathscape was not a promising vista. Our mother was an Irish immigrant, so we understood that another potato famine could strike before dinnertime. We were a family of fundamentalist Catholics, nurtured on graphic stories of martyred saints who sacrificed various body parts—eyeballs, skin—for faith. Even before we were old enough to receive Holy Communion, the priests smeared our foreheads every Ash Wednesday to remind us that we were barreling toward dust. So the drive through this necropolis, with its in-your-face emblems of death, always made me shudder. It had the trappings of a scene from a low-budget horror movie: thousands of headstones flanked the parkway. If traffic slowed, the marbled blur solidified into flocks of angels perched on gravestones adorned with flowers and tiny American flags like the ones we waved on the Fourth of July.

It was Holy Sepulchre Cemetery, today the final resting place of upwards of two hundred thousand departed.[1] But the most unsettling thing about this place was not simply the thousands of graves. Rather, it was the juxtaposition of the thousands of graves and the giant Pabst beer bottle stationed in the sky above them. Erected in 1930, the fifty-five-thousand-gallon Pabst bottle stood on a scaffold on the roof of the Pabst brewery, 185 feet above the highway. Sixty feet from base to bottle cap, the bottle was the same size as the presidential heads at Mount Rushmore National Memorial.[2] I sensed that there was something incongruous about the deified beer bottle fixed in the sky above this sweeping boneyard. Shouldn't that bottle be a statue of Jesus or the Blessed Virgin Mary? I couldn't relax until our car had passed the bottle and put the graveyard behind us.

As a kid, I had absorbed enough iconography to understand that landscape is as much a collection of ideas and values as a book or a law or anything shaped by human imagination, and that when cultures do not physically modify terrain to accommodate their values, they invest it with them nonetheless. In the United States, we have inherited a set of landscape aesthetics—the pastoral, the picturesque, the beautiful, and the sublime—that became wildly popular during the nineteenth century. Aesthetics are ways of thinking and feeling about how things—in this case, the environment—should look, and they reflect as much about a culture's values as they do about its tastes. During the nineteenth century, these aesthetics were frequently wedded to the nation-building credo of Manifest Destiny—a term coined by the editor John O'Sullivan in 1845 to describe the divinely ordained construction of the American empire.[3] O'Sullivan provided a convenient catchphrase for an ideology that predated him, an evolution of what Christopher Columbus and other Old Worldlings had earlier claimed: that God had provided resource-rich terrain for a chosen people to develop while Christianizing the "savages" who stood in their way. We might call such mission-infused versions of these notions about land and beauty *Manifest Destiny aesthetics:* conquer the spectacular if we can, partition the spectacular if we must, use everything else, and romanticize representations of the landscape. With their promise of riches waiting to be plumbed, Manifest Destiny aesthetics often reinforced a class-based system of environmental degradation that sometimes preserved the view.

Why should we care about this now, given the fact that we no longer live in the nineteenth century? Although they continue to evolve across time and geography, the pastoral, the picturesque, the beautiful, and the sublime color the popular media of our times. In contemporary forms, such as advertisements and corporate and political promotional materials, landscape images crafted according to these aesthetics—particularly when combined with allusions to the frontier—often reflect the precepts of the Manifest Destiny era. Because we remain a largely visual culture and because they are transmitted through all types of media, Manifest Destiny aesthetics continue to inform mainstream perceptions of landscape, environmental policies, and representations of national identity, including patriotism, sometimes at great cost.

An Album of American Beauty

During the nineteenth century, the American landscape bloomed beneath the brushes of the Hudson River school painters. Much of their work reflected the still-potent belief of the first European settlers in what Amy DeRogatis, a scholar of American religious history, calls the "moral geography" of American terrain, or the tendency of missionary settlers and emigrants to align the establishment of church-centered communities with "the cultivation and regeneration of souls" and to perceive "moral values inscribed in the landscape."[4] The earliest European settlers had come to the New World looking for wealth and a second Eden.[5] By the mid-nineteenth century, many also saw America as Canaan—a working landscape of seemingly unlimited natural resources that the Creator had provided specifically for them.[6] And much like their heroes of the Old Testament, they came to see the wilderness—particularly mountains—as God's domain. God may have previously resided in heaven, but he had relocated to the United States, where he liked to hang out around farms, forests, and especially mountains.[7] Thomas Cole, now considered our most important early nineteenth-century landscape painter and founder of the Hudson River school, wrote in his 1835 "Essay on American Scenery" that looking at "the loveliness of the verdant fields" and the "sublimity of the lofty mountains" was a spiritual exercise, and it was this "sweet communion" with God that one could bring back from nature walks and disseminate to the democratic community.[8]

The art historian Barbara Novak writes that this "idea of community through nature runs clearly through all aspects of American social life in the first half of the nineteenth century."[9] The "idea of community through nature," or geography, permeated the second half of the nineteenth century as well, but by this time, despite the village picnic overtones, this notion was less benevolent than it sounds.[10] Benedict Anderson argues that nationalism is transmitted through a culture's various media, a list to which the art historian Angela Miller, in the case of the nineteenth-century United States, adds "visual images" of the American landscape. The concepts of the pastoral, the picturesque, the beautiful, and the sublime came to represent the mainstream understanding of the nation's relationship with its landscape and the Christian God.[11] Euro-American residents of the United States began to see themselves as a society increasingly distinct from Europe, and Ralph Waldo Emerson, in his 1841 essay "Self-Reliance," urged Americans to shrug off old ideas and values in favor of original thinking. Yet since the colonial era, Euro-Americans had learned to perceive American terrain according to an inherited set of landscape aesthetics, some of them ancient in origin.

We can trace the pastoral, which romanticizes a simplified, rural life, to early Greek texts and the Bible's Promised Land.[12] Hector St. John de Crèvecoeur's 1782 *Letters from an American Farmer* and Thomas Jefferson's 1785 *Notes on the State of Virginia* updated the pastoral for the new American republic. Written during the nation's seedling years, these texts intertwined American pastoral values with burgeoning American nationalism.[13] Featuring farmhouses, orchards, checkerboard fields, and grazing livestock, American literary and visual pastorals indicated bountifully stocked pantries and projected agrarian independence in a new Promised Land. The American pastoral was God-blessed, and Crèvecoeur argued that cultivation of the soil brought one closer to God—an idea that continues to resonate in contemporary culture with representations of American farm life as wholesome.

Eighteenth- and nineteenth-century British thinkers such as Edmund Burke, Uvedale Price, and the Reverend William Gilpin classified the features of picturesque, beautiful, and sublime land-scapes.[14] These landscape enthusiasts defined the picturesque as that which looked as if it could be framed as a picture—slightly rough and wild, but balanced and not overly rugged. The beautiful was infused with balance and serenity; imagine a serene lake scene or a well-groomed estate. Sublime landscapes revealed the awesome power of God and often featured images of rugged peaks and cliffs, thunder-ing waterfalls, volcanoes, and threatening storms. God showed his might in cliffs and canyons that only mountain goats and bighorn sheep (and only the most daring and highly skilled human climbers) could scale. His muscles rippled the oceans and his fists pounded the shores. He roared in waterfalls and winked in rainbows. Burke envi-sioned sublime nature as the raw handiwork of God, and sublime landscapes remained infused with divinity throughout the nineteenth century and well into the first half of the twentieth; for many nature lovers, they remain so today.

For Burke, sublime landscapes evoked awe infused with fear and a seductive thrill; think of Moses climbing the mountain to receive the Ten Commandments, and then shuddering on the ground at the sight of the back of God. While all of these aesthetics captured the imagination of our nineteenth-century predecessors, it was the sublime that left many of them breathless. Humans seldom figured in paintings of the sublime unless they were Native Americans, whom many nineteenth-century Americans, romanticizing Rousseau's concept of the Noble Savage, considered closer to nature.[15] Those humans who do appear in paintings of the sublime are generally tiny, dwarfed by the power of the Creator revealed in the landscape's dramatic topography.

These aesthetic categories were fluid rather than fixed, and artists frequently featured all of them in a single visual. After the Civil War, many paintings and engravings combined sentiments about the spiritual landscape with the projects of empire, as if the Lord had encoded a blueprint for the nation's development in the features of the landscape.

When Landscape Was in Vogue

The American landscape came flourishingly into vogue in the nineteenth century.[16] William Bartram's 1791 *Travels* was already a classic when Nicholas Biddle published his 1814 *History of the Expedition under the Commands of Captains Lewis and Clark.* The expedition journals would not be published in their entirety until 1904–1905, yet Biddle's popular version of the journals of Meriwether Lewis and William Clark struck a timely chord; it was a nationalistic, "action and adventure" narrative that lionized the explorers, degraded Native peoples, and endorsed the projects of empire-building.[17] In 1831, Mount Auburn, the first American rural cemetery and a precursor to Holy Sepulchre Cemetery, opened outside of Boston. Our first public parks, rural cemeteries were considered "pleasure grounds for the general public," but they were also picturesque places for exercise and meditation where, surrounded by nature and reminders of mortality, one might humbly contemplate spiritual matters.[18] Washington Irving, already famous for his 1819 *Sketch Book* and other writings, published *A Tour on the Prairies* in 1835, *Astoria; or, Enterprise Beyond the Rocky Mountains* in 1836, and *Adventures of Captain Bonneville* in 1837.

Thomas Jefferson and Hector St. John de Crèvecoeur, championing the agrarian republic, had looked to American soil, but it was Thomas Cole who raised the nation's gaze and fixed it on American scenery. "The most distinctive, and perhaps the most impressive, characteristic of American scenery is its wildness," the artist pronounced in 1835, igniting a bias for wild scenery that would influence painted and printed representations of the American landscape from that moment forward.[19] In 1836, Cole elevated American landscape painting to a high art with his *View from Mt. Holyoke after a Thunderstorm—The Oxbow* and his five-canvas cautionary tale, *The Course of Empire.* That same year, Ralph Waldo Emerson delivered his essay "Nature," in which he documented his metamorphosis into a "transparent eyeball," an apt—if rather inelegant—description of his communion with the divine Over-soul during his walks in the woods around his home in Concord, Massachusetts. It is no coincidence that the transcendentalist philosopher described himself as an eyeball and believed he could withstand anything except blindness.[20] Beyond the obvious difficulties of a sightless life, blindness would have deprived him of the view that transported him to his meditative states. "The health of the eye seems to demand a horizon," he insisted, forecasting the importance that gazing upon landscape would assume throughout the expansionist era.[21] Margaret Fuller would describe farther horizons in her 1843 *Summer on the Lakes.*

In the following decades, painters such as Frederic Edwin Church, Asher B. Durand, Thomas Moran, and Albert Bierstadt, to name only a few, added their own visions of the American landscape to the album that Cole had started.[22] In his editorial columns in the *Horticulturist* and in the 1849 edition of his *Treatise on the Theory and Practice of Landscape Gardening, Adapted to North America*, the landscape designer A. J. Downing advised readers to fashion their properties according to the picturesque and the beautiful, and to construct homes with a "republican simplicity" befitting the United States.[23]

Susan Fenimore Cooper published her observations of rural New York in the first edition of *Rural Hours* in 1850. G. P. Putnam's 1852 *Home Book of the Picturesque* included essays and engravings by some of the nation's most prominent literary and visual artists. Henry David Thoreau built his reconstituted hut at Walden Pond in the 1840s and published his intimate observations of nature and society in the quietly received *Walden* in 1854.

By the late 1850s, more than forty horticultural societies had formed in response to what the historian Tamara Plakins Thornton calls "an explosion of popular interest in horticulture."[24] Frederick Law Olmsted designed some of the nation's most important parks, including New York City's Central Park (undertaken with Calvert Vaux) in 1858. John Muir, the naturalist who would later found the Sierra Club, began trekking across the United States in 1867. Between 1855 and 1891, Walt Whitman published various editions of *Leaves of Grass*, with expansive lines of verse that reflected the "democratic vistas" of the American continent. Although her work would not become widely available to the general public until the twentieth century, Emily Dickinson blossomed into a passionate backyard naturalist who penned more than 1,700 poems between 1850 and 1885. In 1871, John Burroughs published his observations of the natural world in *Wake-Robin;* multiple works would follow. Appleton's published volumes of the enormously popular *Picturesque America* between 1872 and 1874. And in 1872, Congress designated Yellowstone our first National Park.

All of these developments occurred within the context of the explorations and geological surveys of the continent; waves of immigration to the United States and emigration across the continent; the laying of the railroads and the expansion of the American industrial empire to the Pacific Coast; the Indian removals and the Mexican, Civil, and Indian wars; and the passage of the Homestead Act of 1862, the Mining Act of 1866, the Timber Culture Act of 1873, and the Dawes Severalty Act of 1887, all of which exerted a dramatic impact on the landscape and the people who inhabited it. The nation was engraving the story of its course of empire onto the land.

Picturing American Nationalism

Uncomfortable with their image as Europe's philistine half-siblings, nineteenth-century Euro-Americans turned to their continent's geography to secure a sense of national identity. Their country lacked the Old World's ancient ruins, so they took pride in a landscape that included such spectacular features as the Rocky Mountains, Niagara Falls, the otherworldly terrain of Yellowstone, and the colossal trees of California and the Northwest. They were awed by the sheer breadth of the continent and dazzled by the bounty of its forests, lakes, rivers, and mineral resources. "Nationalists found in the country's geographic scale the measure of their own cultural merit," writes Angela Miller.[25] Landscape painters documented geological wonders and expansionist progress for a public hungry for romantic images of the almost mythical American landscape.

The landscape paintings that hung on the walls of eastern and—by the second half of the nineteenth century—western galleries and museums, and the engravings that graced the pages of magazines like *Scribner's*, *Century*, and *Appleton's Journal* and books such as the popular thousand-page parlor table book *Picturesque America* taught the American public to picture their nation in very particular ways. Americans learned to look for rolling slopes and herds of domesticated animals grazing in the lush fields of pastoral vistas. They learned to note, in the foreground of a canvas, the dead snag or fallen tree or stump—America's version of ancient European ruins, or perhaps icons signifying the collapse of the empire of wilderness beneath the new civilization. They learned to appreciate the slightly rough, but not terribly rugged, landscapes of the picturesque and the serenity and balance of the beautiful. And they learned to recognize the fingerprints of God in the serrated peaks of the Far West. After the Civil War, the geography portrayed in landscape images would provide the common ground over which the nation could reunite and heal.[26] The historian Sue Rainey writes that by the mid-1870s, *Picturesque America*, which included intricate engravings of scenic places across the nation, "promoted and reinforced a resurgence of nationalism that was rooted in the [nation's physical geography] rather than in the institutions of democracy."[27] The geographic features of the American landscape, depicted according to the pastoral, the picturesque, the beautiful, and the sublime, became symbols of the dominant national community.

The Iconography of Progress

Thomas Cole expressed serious reservations about the development that was transforming the northeastern landscape. Although he did not dismiss the importance of cultivation and industry, he decried what he called the "meager utilitarianism" of the times.[28] But if Cole was troubled by encroaching industrialism, many of his successors expressed far less skepticism. Emerson wrote in 1836 that "Nature always wears the colors of the spirit."[29] It's a remark that I think of as the Emersonian mirror—a tendency to project one's own feelings and preconceived notions onto one's surroundings. Emerson cultivated the "habit of attention" that enabled him to transcend that mirror to glimpse what truths he might beyond its surface.[30] But many of Emerson's compatriots were more preoccupied with the material resources latent in nature. After Cole's death in 1848, nature—as it was represented in much of the ensuing landscape imagery—often "wore the colors of the spirit" of American expansionism. Several of the later artists, including Albert Bierstadt and Thomas Moran, whose work was often commissioned by railroad companies seeking to encourage cross-continental tourism, emigration, and land sales, created works chronicling the triumphs of Manifest Destiny.

Many of these later paintings told more or less the same story. Angela Miller notes that by the mid-nineteenth century, "landscape art . . . was a cultural endeavor directed at consolidating a middle-class social identity bound up with the civilizing mission."[31] Emblems of this "mission" were paintings such as Emanuel Gottlieb Leutze's 1861 *Westward the Course of Empire Takes Its Way.* Featuring a wagon train complete with a pioneer version of Joseph, Mary, and the infant Jesus, the painting casts emigrants in the roles of pilgrims on a holy quest and the story of emigration as sacred scripture. John Gast's 1872 *American Progress* portrays a Goddess of Liberty, who could have been a nineteenth-century prototype of a Barbie doll, trailing telegraph wires westward across the continent as Native Americans and bison run literally out of the picture.

Miller argues that "American landscapes at the time were moral exhortations to restraint and discipline of both self and society. . . . They associated formal hierarchies, economic colonization, and the domestication of wilderness with the parallel project of containing and socializing the passions of the republic's citizens."[32] Post–Civil War landscape images often functioned much like corporate and political advertisements, endorsing a worldview that included expansionist policies laden with complex social and environmental implications. These images depict landscapes of promise and celebrate the frontier that Frederick Jackson Turner, in his 1893 speech, "The Significance of the Frontier in American History," would claim had molded the "coarseness and strength," practical idealism, and civic-mindedness of American character. The ornate frames in which paintings were mounted presented a vision of nature that was "contained." It was this mastery of the physical world that was largely the point; the frames signified civilization's domestication of the wild continent.[33] The images do not depict the genocide of Native Americans or the environmental degradation—such as the massive logging around Lake Tahoe that accompanied Nevada's Comstock silver boom—that underscored such visions of progress. Those stories remained outside the frame.

The Sacred Sublime

Some who had reservations about the nation's particular course of empire may have been placated by Thomas Moran's *Mountain of the Holy Cross* (fig. 1). Moran's painting depicts a geological phenomenon—two bisecting, snow-filled fissures that form a cross on the face of a mountain deep in the Colorado Rockies. W. H. Goetzmann and W. N. Goetzmann note that "the *Mountain of the Holy Cross* became an archetypal image of a Christian nation, an outward sign that God himself had blessed the westward course of empire."[34] Moran's *Mountain of the Holy Cross* may have been a prototypical Manifest

FIGURE 1: Thomas Moran, *The Mountain of the Holy Cross*, 1875. Museum of the American West, Autry National Center of the American West, Los Angeles, 91.221.49.

Destiny painting, but it tapped into American notions of nature—
especially sublime mountain ranges—as God's cathedral. The sublime was conscripted, Barbara Novak writes, "into a religious, moral,
and frequently nationalist concept of nature, contributing to the rhetorical screen under which the aggressive conquest of the country
could be accomplished."[35] As landscape imagery gained prominence
in the fine arts and became linked with expansionist mythology, so it
permeated mass culture, where print media disseminated intertwined
notions of landscape and nation, solidifying the myth and carrying it
forward into our times.

By contrast, naturalist John Muir saw the artistry of God, uncou-
pled from visions of empire, in the sublime. In *My First Summer in
the Sierra*, Muir, well-indoctrinated in nineteenth-century landscape
aesthetics, found the Lord everywhere in Yosemite, and Muir's God
would have made a trendy landscape painter. Perhaps because he
had earlier been temporarily blinded in a workplace accident, Muir
exulted in the pleasures of vision. His book reads like an aesthetic
guide to the landscape of Yosemite, noting where the scenery is sublime, beautiful, or picturesque. At least one of those terms appears
on perhaps four-fifths of the book's pages, and Muir almost always
associates them with the sacred. For John Muir, nature is "sculptor," "architect," "painter," "gardener," and the text of "God's divine
manuscript."[36] And the ground of Yosemite is God's temple. Penned
in 1869 and published in 1911, *My First Summer in the Sierra* is one of
the benchmark texts of American environmental writing, so it is easy
to understand why environmentalists ever since Muir have been so
taken with scenery, and why, for many, sublime landscapes are sacred
ground.

If they could have seen it, Thomas Cole and John Muir would have
been horrified by twentieth-century New Jersey's monumental Pabst
beer bottle stationed above the thousands of graves at Holy Sepulchre
Cemetery. Holy Sepulchre, part of the trend toward establishing
rural cemeteries as spaces for outdoor meditation and recreation,
opened in 1859. The bottle was raised to promote a brand of ginger
ale in 1930 and it became a *beer* bottle when the brewery took over in
1945.[37] The juxtaposition of the bottle and the burial ground seemed
incongruous to me as a kid only because I had a child's education in

the iconography of power. Since my family attended a church inhabited by gargantuan statues that hung above the crowd of worshipers,
I assumed that anything that big—and elevated above a city of dead
people—should have represented God or one of his most favored
assistants. In its own way, the bottle did represent something godlike. With its massive dimensions, the bottle exemplified what Leo
Marx calls the "technological sublime," indicative of the power of the
engineering that, in this case, facilitated mass production of Pabst
Blue Ribbon Beer.[38] The bottle and the boneyard reflected evolving
cultural views about landscape and power. They were incompatible
only according to the aesthetic theories that both Thomas Cole and
John Muir had embraced and that still informed much of the media
to which I had been exposed.

The Nineteenth Century in the Twenty-First

Contemporary versions of the aesthetics that saturated this nation
in the nineteenth century continue to show up everywhere in popular culture. We find them in the calendars hanging on our office
and kitchen walls, on the pages of home and garden magazines, on
television programs, in virtual space, and in the ads that form the
background images of our daily lives. Ads for consumer commodities frequently feature images of the American landscape. The communications theorist Julia Corbett finds that "nature-as-backdrop
ads portray an anthropocentric, narcissistic relationship to the biotic community and focus on the environment's utility and benefit
to humans," values that come straight out of the Manifest Destiny
era.[39] Political candidates often seek to bolster their images with an
injection of the pastoral, and all its attendant values, by appearing in
shirtsleeves beside a John Deere tractor and red barn at a rural town
meeting.

In his 1893 speech about the frontier's influence on American
character, Frederick Jackson Turner, who came of age in the second
half of the nineteenth century and was perhaps influenced by the
sentiments behind so many of the landscape images that had formed
the panorama of his own early life, conjoined the ideas of democ-

racy and the frontier. This union formed what Richard Slotkin calls the "Myth of the Frontier," an evolving narrative marked by "poetic construction[s] . . . capable of evoking a complex system of historical associations by a single image or phrase. For an American," Slotkin continues, "allusions to 'the Frontier' . . . evoke an implicit understanding of the entire historical scenario that belongs to [an] event and of the complex interpretive tradition that has developed around it."[40] As the political scientist John Tirman observes, frontier mythology encompasses "the limitless possibilities of the American dream, the expansion of American values, the national effort to tame faraway places, the promise of a bounty just over the horizon, and the essential virtue of the American people who explore and settle these frontiers," and it continues to inform contemporary discourse. Tirman adds that the 2008 "run for the White House"—featuring John McCain as a warrior hero, Sarah Palin as a "frontier mother who hunts," and Barack Obama as a latter-day Abraham Lincoln given to alluding to "new frontiersman" John F. Kennedy—"recycled the frontier myth with scarcely a nod to its growing irrelevance."[41]

Landscape representations may trade on or argue with frontier mythology. But contemporary images of frontier or western iconography are frequently designed to evoke the dominant strain of this nation's creation stories, recalling the cross-continental treks of the pioneers and the expansion of the American empire celebrated in much of the nineteenth-century's landscape art.

In the United States, landscape aesthetics are so deeply embedded in the "poetic constructions" of national mythology that they continue to offer us a way of identifying ourselves as Americans. Robert F. Kennedy Jr., echoing Frederick Jackson Turner, remarks that "our landscapes connect us to our history" and that "they are the source of our character as a people."[42] And from the time we are very small, we are taught to associate particular landscape images with patriotism. Along with "The Itsy-Bitsy Spider" and "I'm a Little Teapot," preschoolers learn the lyrics to patriotic songs. "God Bless America" links God with a geography that specifically includes mountains and prairies. "My Country 'Tis of Thee" fuses liberty, war, pride, and mountainsides. "America the Beautiful," which includes the aesthetic of the beautiful in its title, celebrates the Jeffersonian pastoral in "amber waves of grain" and "the fruited plain" and the sublime in "purple mountain majesties." The song connects all of these aesthetics to topography, democracy, and a history of development in lyrics that recall the arrival of the pilgrims and the frontier-bound treks of the pioneers. "God Bless America" and "America the Beautiful" are hymns of the American civil religion. They venerate coast-to-coast geography and imply reverence for the features of American terrain.

Across the Continent

I moved to Montana because I had fallen in love with a map and a myth. In my New York City apartment, I had found the West in catalogs of Thomas Moran's paintings, and I had explored Montana in my atlas and in Richard Hugo's poetry about inarticulate people in eloquent towns. I felt caged by my urban environment and imagined that the West's expansive geography would open some of the cramped channels in my mind and heart. This sentiment was a holdover from our expansionist era; Yi-Fu Tuan writes that nineteenth-century Americans associated wild nature with liberty, a "feeling," he says, that "lingers into our time."[43] I never surmised that I had been influenced by the western-themed films and television programs that had formed part of the backdrop of my childhood. As a kid I was crabgrass lawn, not shortgrass prairie, and I hated westerns, with their gun-slinging men and endless shoot-outs. I never imagined that I was responding to the white noise of American aesthetic discourse.

I worked for a time scrubbing pots and baking bread at a guest ranch outside of Choteau, along Montana's Rocky Mountain front. This was mythical American terrain, where the plains smacked into the Rocky Mountains. Choteau was surrounded by farms with silos that gleamed in the sunlight and ranchland speckled with grazing cattle. Pheasants hid in tall grasses and pronghorns vaulted along the hummocks that bordered the back roads. The work was hard, with heavy lifting that tore at my back, but I knew I was tremendously fortunate to work among kind people in a beautiful place. And I had lived so long in New York City that I felt I could not get enough of Montana. Often, in the evenings after work, I drove through the countryside, marveling at a landscape that was ribbed with dinosaur bones and shaped by glacial fingers. One evening I headed along my favorite stretch of road toward the Sun River valley town of Augusta. To the east, rain pummeled the plains. Forks of lightning shredded the sky and a rainbow sloped from slate thunderheads down to the earth. To the west, backlit by a brilliant red sunset, the serrated Rockies glowed violet blue. A trio of cowboy brothers on horseback worked their cattle just inside a fenceline beside the road. Here pastoral, picturesque, beautiful, and sublime frontier collided in one vast panorama. I was stoned on scenery, drunk on American mythology. I thought I had never before seen anything quite so beautiful or so quintessentially American.

It was the stuff of myths, movies, and calendar photos. But of course, it was only the surface. Subsurface, a nuclear missile nestled in its silo at the edge of a gravel road just a short distance from the ranch where I worked. Pockets of nuclear missile silos, discreet squares of earth cordoned off by chain-link fencing, punctured that Montana landscape. And it was from the Sun River valley that I so admired, and from Missoula, my home for two years and where the United States had imprisoned Italian nationals, Japanese residents, and Americans of Japanese ancestry during World War II, that troops of U.S. cavalry and infantry had been dispatched to slaughter the Nez Perce in the battles of Big Hole and Bearpaw in 1877.[44] Those battles, and the image of Lewis and Clark crossing the territory with Clark's slave, York, and with Sacagawea, a nursing Shoshone mother and their conscripted guide, were as quintessentially American as the panorama I saw on the road to Augusta. So, too, was Butte's Berkeley Pit, the nation's largest Superfund site, barely 165 miles from that scenic stretch of road.

"Social psychologists, cultural geographers, and cultural anthropologists have established the profound degree to which culture influences perception," writes the aesthetics theorist Arnold Berleant.[45] I eventually realized that I was so thoroughly indoctrinated in the landscape aesthetics of national iconography that I had spent all of my life looking at the landscape as I had been trained to see it—through a lens distorted by Manifest Destiny aesthetics and twentieth-century visions of patriotism. The first time I visited Wounded Knee, on South Dakota's Pine Ridge Reservation, I sought out the cemetery where 144 Sioux were buried in a mass grave after they were massacred by the U.S. Cavalry on December 29, 1890. The cemetery is small, yet I failed to recognize the marker of the mass grave even though I stood directly in front of it for some time. In the privileged world in which I grew up, national monuments tend to be big and glossy and set at some remove from their surroundings. At Wounded Knee, where the tribe has refused government help—or interference—the local folks continue to bury their dead in humble graves beside the modestly marked mass grave.

Seventy miles from Wounded Knee, the massive sculptures of Presidents Washington, Jefferson, Lincoln, and Roosevelt stare out from Mount Rushmore National Memorial. Most of us know Mount Rushmore from the barrage of tacky ads to which we are treated annually as marketers promote their Presidents Day sales. The rest of the year, images of the monument are used to hawk everything from calcium supplements to sleep aids to homeland security policy. The monument's official moniker is "America's Shrine of Democracy," but it is located on land that was deeded to the Sioux in the treaty of 1868, and then taken back after George Armstrong Custer confirmed reports of gold in the Black Hills. The Black Hills are sacred to the Sioux, and the sculptures are carved into a mountain that, by definition, we would label sublime—God's handiwork, and no place for puny humans except as interlopers.

The heads at Mount Rushmore measure sixty feet from chin to crown, and the monument faces in the general direction of the Pine Ridge Reservation, which is one of the poorest communities in the United States.[46] When their sculptor, Gutzon Borglum, blasted the faces of four presidents onto the mountain, he elevated the human to the sublime, rendering a select group of men as godlike patriarchs and the dominant story of American expansionism as sacred text inscribed directly on the landscape. We may call Mount Rushmore our Shrine of Democracy, but it would be more accurately labeled a monument to white patriarchy and imperialism. If the heads of the presidents had eyes that could see and ears that could hear, they would have remarked the explosions of missiles in the nearby Badlands Bombing Range, where over three hundred thousand acres of Pine Ridge Reservation land were used as air-to-air and air-to-ground gunnery ranges from 1942 through 1945.[47]

Reading Dirt

Emerson sought revelation by reading the book of nature. When I moved out West, landscapes became my texts. During the years I lived in Montana and after I made my home in Nevada, I began to look directly on the ground for traces of the history I had never been taught in school. I sought out the stories of the places I saw, scrutinized the language used to tell those stories, and examined the ways that mainstream culture uses our storied landscape to bolster a particular vision of American identity. I analyzed policy documents and artifacts of material culture, such as advertisements, postcards, and calendars, to see how they drew upon the rhetoric—verbal or visual—of patriotism. I examined some of the inconsistencies in the landscape aesthetics of national myths, such as the practice of veiling clearcuts behind a thin screen of trees so that passersby see the picturesque facade rather than the devastated ecosystem behind it. I dissected the mythology I found in so many contemporary landscape representations. And I began to contrast these representations with what was actually on the ground.

I found that many contemporary landscape images are composed according to the aesthetic styles, with all their accompanying nationalistic ideologies, of the Manifest Destiny era. As many nineteenth-century images worked to promote nationalism, contemporary images in these styles—with the notable exception of those used by conservation organizations—often serve as *patriotic totems* that encourage acceptance of government policies, business practices, and cultural standards by dressing them up in the aesthetic trappings of the nation-building era. They are often couched in rhetoric that draws on frontier mythology and patriotism. Like the expansionist era paintings and engravings, they portray nature as "contained" and they leave the bloodshed and environmental degradation out of the frame.

Such framing is at least as old as the nineteenth-century government policies that partitioned landscapes into the utilitarian and the
scenic; with the passage of the Wilderness Act in 1964, we added
the partition of the wild. Examining contemporary American perceptions of wilderness, the historian William Cronon writes that "as
we gaze into the mirror it holds up for us, we too easily imagine that
what we behold is Nature when in fact we see the reflection of our
own unexamined longings and desires."[48] Cronon also notes that in
the nineteenth century, the sublime and the frontier "converged to
remake wilderness in their own image, freighting it with moral values
and cultural symbols that it carries to this day."[49]

I believe that we might well extend Cronon's argument beyond wilderness to all landscapes. I contend that all of nature functions as the
Emersonian mirror onto which the policy-makers project and enact
"the colors of the spirit" of the times. And if wilderness is invested
with "moral geography" and "cultural values," then so are scenic and
utilitarian landscapes. Our aesthetic traditions have motivated us to
preserve some spectacular places. But we sometimes value scenic
landscapes above most others, including those in which we live and
work.[50] As the literary and cultural theorist Lawrence Buell writes,
landscape imagery may persuade us "to cordon off scenery into pretty
ghettoes."[51] This can conceivably foster what the ecocritic Cheryll
Glotfelty calls a "placist" attitude—a form of discrimination against
places which Glotfelty aligns with racism.[52] One of the potential consequences of "placism" is that conventionally scenic landscapes may be
protected, while others may be devalued and even abused. Yet sometimes even scenery is not enough. Looks count in Manifest Destiny
aesthetics, but they often count more in corporate and political ads
than they do on the ground.

In the Name of the Bomb

The Wasteland's Atomic Bloom

And what rough beast, its hour come round at last,
Slouches towards Bethlehem to be born?

William Butler Yeats, "The Second Coming"

It was morning in Nevada, President George W. Bush was in office, and all over America's party capital, people were waking up or turning in. At 7:00 a.m., the boomtown's glitter had dimmed to grit. A couple of miles off the Las Vegas strip, where a Statue of Liberty stands a few city blocks from a Cinderella castle fringed with palm trees, a security checkpoint siphoned me onto a bus outside the Atomic Testing Museum. I was making a pilgrimage to another kind of boomtown, a mission I could undertake because the Department of Energy (DOE) had determined that I am not a threat to national security. Several times each year, the DOE contractor Bechtel Corporation ferries two busloads of tourists, shorn of cameras, laptops, tape recorders, and cell phones, sixty-five miles north of Las Vegas to the Nevada Test Site, which served as the United States' nuclear proving ground from 1950 until the nuclear weapons testing moratorium of 1992. I was traveling to this bombscape because I wanted to try to make sense of the violence enacted here in the name of peace and of the terror that had governed my cold war childhood clear across the continent. What I found was a pocked and irradiated landscape littered with war trash and celebrated in language that fused nostalgia, patriotism, and a burly sexuality.

Arrivals

My mother was a lovesick Irish immigrant on the lam from a belly-up romance when she disembarked at New York Harbor in 1955. She was not yet an American citizen when my parents married the following year, so the authorities considered her a security risk and they fired my father from his job with the CIA three days after my parents' wedding. My mother delivered me in our nation's capital in February of 1959, just days before Mattel introduced Barbie, a doll that codified in plastic the ideals against which the culture had long measured women: long-limbed and slender with incongruously large breasts, feet eternally arched for stiletto heels, lipsticked mouth perpetually closed but for a demure smile. Some of Barbie's later incarnations would be engineered by a former weapons designer whose house, claimed his ex-wife, included a "torture chamber," but the doll's debut followed a market research study that tested the contents of an eloquent toy box of cold war playthings: a Barbie prototype, guns, rockets, and holsters.[1] It was an era when the power of the white patriarchy was challenged by iconoclasts like Allen Ginsberg, who howled to the music of "the hydrogen jukebox" from the urban wilderness of San Francisco or New York City. Although none of the 1,053 nuclear bombs detonated by the Atomic Energy Commission, reincarnated in 1977 as the Department of Energy, was exploded in 1959, it was, like all cold war years, just one of many years of the Bomb.[2]

The first year of the Bomb commenced with a nuclear dawn on July 16, 1945, when physicist Robert Oppenheimer and his colleagues hatched the egg-shaped Trinity, the world's first atomic bomb, near Alamogordo in the New Mexican desert. For humankind, the birth of the Bomb—the destroyer—was perhaps as significant as the birth of Jesus Christ—to believers, the savior. The name of Oppenheimer's bomb strikes me like a blow to the gut. *Trinity.* To someone raised as a Catholic kid, reluctantly or not, Trinity signifies a three-faced patriarch of Father, Son, and Spirit—a force of impalpable love conjoined with formidable anger, and of authority unquestionable except by those who would court eternal damnation.

Trinity. Had technology replaced God in this nation that has, since the pilgrims first staggered ashore, liked to portray itself as a Christian country? Perhaps the atom—neutrons, protons, electrons weighty as spirits—had become our new Trinity. In *Flash Effect*, David Tietge observes that during the "atomic age . . . science in many ways functioned as a surrogate for religion," and that Trinity symbolized the "convergence" of science and religion. Trinity gave us "the immediate understanding that human beings had captured a godlike power that required equally godlike accountability," Tietge writes.[3] Oppenheimer understood that he had duplicated the sexton's keys to an unholy tabernacle, and that humankind possessed neither the knowledge nor the grace to manage what his team had released. A few weeks later, seventy thousand people died instantly when our second nuclear bomb, Little Boy, fell on Hiroshima.[4] Another one hundred thousand Japanese perished when a Humpty Dumpty–shaped twin of Trinity, called Fat Man by its creators in honor of Winston Churchill, exploded over Nagasaki.[5] Thousands died more slowly from radiation sickness and burns, and eighty thousand Nagasaki survivors were left homeless.[6] Perhaps some of Nagasaki's large Catholic population noted the cruciform figure of the plane just before it bombed them. "Both detonations were intended to end World War II as quickly as possible," the DOE advises in *United States Nuclear Tests*, casting the bombs as tools of peace.[7]

The sixteenth-century explorer Ferdinand Magellan named the Pacific Ocean for its peaceful waters, but within a year of the bombings of Japan, the United States was exploding nuclear bombs in the South Pacific's Marshall Islands.[8] In *Bombs in the Backyard*, A. Costandina Titus notes that in the spring of 1946, the 167 residents of Bikini Atoll were removed to a small island, where the fishing was poor, at some distance from their ancestral home. The Bikinians agreed to go, but the operation recalls the nineteenth century's Indian removals, when the United States military forced indigenous peoples—in the name of American progress—off their homelands, often onto marginal terrain that could not sustain them. On July 1, 1946, the United States dropped the first of many nuclear bombs on Bikini Atoll.[9] At a Paris fashion show four days later, French designer Louis Reard ignited another kind of firestorm when he introduced the world's skimpiest swimsuit to date—scraps of newspaper-printed fabric that he called the *bikini* because, he said, like the nuclear bomb, it represented "the ultimate."[10] Photos from the show portray a bikini-clad model holding the matchbox into which she could fold her swimsuit when she removed it. Subsequent South Pacific bombings sickened both neighboring islanders and U.S. military servicemen. "By 1963, when the Partial Test Ban Treaty went into effect," Titus observes, "106 nuclear weapons, including the hydrogen bomb, had been detonated in the South Pacific."[11] Congress has since made financial reparations to the islanders, but as of this writing, the Bikinians have been unable to return home because their home is no longer fit for habitation.[12]

Pacific testing posed both public relations and security hazards for the United States, and when the Soviet Union detonated its first nuclear weapon in 1949, our government began to search for a continental test site—an operation given the homespun-as-apple-pie name of Project Nutmeg.[13] The ensuing arms race between the United States and the Soviet Union, based on the principle of Mutually Assured Destruction, now seems perfectly aligned with its acronym: MAD. I recall none of the religious rhetoric that accompanied representations of the Bomb. What I remember is the fear.

"Worthless" Land

I grew up in New Jersey and later lived in Nevada—two states that are geographically antithetical but that share several common problems, including the strains that overpopulation imposes on limited resources, development encroaching on ever-shrinking open spaces and wildlife habitats, and environments perpetually threatened by industrial waste. Cheryll Glotfelty remarks that both New Jersey and Nevada fall victim to what she calls "place bashing."[14] Just admitting that I grew up in New Jersey usually draws a disparaging remark of some kind, since the Garden State is also associated with excessive use of hairspray and a discordant accent entirely unlike the nasal vowels of the real thing. Both New Jersey and Nevada are frequently labeled either the armpit or the asshole of the nation; accordingly, Nevada's Yucca Mountain has long been targeted as the favored site for the nation's nuclear toilet.[15]

The pervasive attitude toward Nevada's desert landscape isn't surprising. When it comes to landscape, most Americans like their greens, and the pastoral, the picturesque, the beautiful, and the sublime map the topography of much of our media. Early explorers and emigrants, accustomed to European aesthetics and eastern foliage, were bewildered by the Great Basin.[16] Glotfelty notes that the forty-mile journey across Nevada's Carson Sink "was the most dreaded leg of the odyssey."[17] Forty is a number that early Christian emigrants would have understood as biblical, since the Hebrews stumbled through the desert for forty years before they finally reached the Promised Land. One psalmist observed, "Some wandered in desert wastelands, finding no way to a city where they could settle. They were hungry and thirsty and their lives ebbed away."[18] According to the Gospel of Luke, Jesus took to the desert for forty days at a stretch for bouts of fasting and praying, and it was in the desert that Jesus encountered the devil.[19] Emigrants found the crossing of the austere Nevada landscape appropriately unholy, and they may have recalled that one of Christ's greatest agonies during his torture and crucifixion was his excessive thirst. Since Nevada lacked the rainfall and greenery to render it picturesque, beautiful, or even benevolently sublime, it was simply a "waste." The spirit of the times was overwhelmingly utilitarian. Even Ralph Waldo Emerson, an eloquent lover of nature, subscribed to a "doctrine of Use." "Nature . . . is made to serve," he asserted. "It receives the dominion of man as meekly as the ass on which the Saviour rode. It offers all its kingdoms to man as the raw material which he may mould into what is useful."[20] That utilitarian spirit has not diminished greatly in the last 150 years. Think about the last time you drove across the border of a national forest. What did the sign say? "Land of Many Uses."

Until prospectors unearthed a mother lode of silver, the early emigrants could discern little use in country like Nevada. And they had a point; Nevada's average annual precipitation is a mere eight inches, and a land with so little water was not meant to support large agricultural operations or frenetic cities. But by 1894, the State Bureau of Immigration, in *Nevada and Her Resources*, was trying to convince prospective residents that Nevada could, with irrigation, become the pastoral promised land of Thomas Jefferson's dreams. "Someday artificial lakes will dot the landscape, preserving the floods [of the Truckee, Carson, and Walker Rivers] to make fruitful thousands of acres of land now worthless," the boosters promised.[21] They had plenty of biblical precedent, since several of the Psalms promised that the Lord would make streams flow in the desert. "The grasslands of the desert overflow," rejoiced one psalmist. "The meadows are covered with flocks and the valleys are mantled with grain."[22]

Nevada's Manifest Destiny

Nevada's 1894 boosters succeeded, and state residents have since reinvented parts of the desert as pastoral farming and range land, and parts as picturesque, if only to those who live here. And whether or not God had anything to do with it, we have made streams flow in the desert. In the wake of the nuclear testing moratorium of 1992, the DOE, reinventing the Nevada Test Site as a place where corporations could store their nastiest waste and conduct their most hazardous experiments, promised potential clients that the Test Site's wells could provide up to nine million gallons of water per day.[23] The DOE has even attempted to recast the Nevada Test Site as picturesque, perhaps because, to Americans, that aesthetic is almost universally appealing. The first page of *The Nevada Test Site: A National Experimental Center,* a 1994 DOE brochure, features a photo of a bristlecone pine and the Test Site's Rainier Mesa and Stockade Wash (fig. 2). The photo is composed much like a Hudson River school painting; the main difference is that this is a shot of a desert landscape rather than a lush northeastern one. The image includes many of the tropes of nineteenth-century landscapes: a prominent bristlecone pine standing in for the snag in the left foreground; sloping mountains; roads that meander like rivers; a cloud-laden sky.

What the picture doesn't reveal is that Rainier Mesa's surface conceals a network of sixteen tunnels where roughly seventy underground bombs were exploded for about $60 million a pop.[24] Costandina Titus notes that Rainier Mesa was critical to the Reagan administration's defense operations. In 1984, Titus reports, "DOE spokesman Dave Miller told the press that testing at Rainier Mesa . . . was crucial to the United States 'Star Wars' defense; he claimed that this was where 'America is learning how to knock out Russian satellites . . . [and] how the MX and other missiles could take the Soviets' best shot.'"[25] Like

Figure 2: *Bristlecone Pine, Rainier Mesa, and Stockade Wash*. Photo courtesy of National Nuclear Security Administration / Nevada Site Office.

the nineteenth century's Manifest Destiny landscapes, the photo prettifies the story of who and what had suffered, or would suffer, beneath the ideologies of conquest and defense.

Although many who live in Nevada appreciate its singular beauty, much of the rest of the country prefers a more traditional vision of the picturesque, and therefore views Nevada as a wasteland rendered tolerable by its tax laws and air-conditioned casinos. So it's not terribly surprising that Nevada became home to the nation's nuclear *proving ground*—an expression that raises the question of why we must prove ourselves, or anything, against the environment—or that lawmakers have often viewed Nevada's Yucca Mountain as a potential national nuclear dumpsite. In 1940, our government figured out how to make the wasteland useful when it set aside the Las Vegas Bombing and

Gunnery Range, which, a decade later, became Nellis Air Force Base. In December 1950, President Truman designated a portion of this land as the Nevada Proving Ground, and within six weeks the Atomic Energy Commission was dropping nuclear bombs on Nevada. According to the DOE, the first shot, Able, left a "violet afterglow."[26] "It's exciting to think that the submarginal land of the proving ground is furthering science and helping national defense," Nevada governor Charles Russell enthused the following year. "We had long ago written off that terrain as wasteland, and today it's blooming with atoms."[27]

One year after Governor Russell's remarks and ten years after American men had begun to refer to beautiful women as "bomb-shells," Las Vegas showgirl Candyce King was named Nevada's first Miss Atomic Blast. The caption beneath King's newspaper photos read, "Radiating loveliness instead of deadly atomic particles, Candyce King, actress appearing at [the] Last Frontier Hotel in Las Vegas, Nevada, dazzled U.S. Marines who participated in recent atomic maneuvers at Yucca Flats. They bestowed on her the title of 'Miss Atomic Blast,' finding her as awe-inspiring, in another way, as was the 'Big Bang.'"[28] Nevada designated its last and most famous atomic beauty, now called Miss Atomic Bomb, in 1957. Showgirl Lee Merlin posed for photos with her arms stretched skyward, the cottony form of a mushroom cloud tacked across her torso.[29]

Las Vegas residents and visitors could watch the nuclear fireworks until 1963, when the Limited Test Ban Treaty put a halt to atmospheric testing.[30] Subsequent bombs were exploded underground—jolts that created massive plumes of rock and dirt shaped like ephemeral warts that immediately collapsed into craters as the earth around the Bomb vaporized.

The Lexicon of Violence

Our bus rumbled along Highway 95 toward the Test Site, and I marveled at the mountains, starkly beautiful in the early morning light. Our tour guide was a fifty-year veteran of the DOE whom I will call Howie. A congenial man, Howie was seventy-one, retired, and so good-natured that he might have been born grinning. Howie played

DOE videos on the small screen at the front of the bus, and as we drove I watched mushroom clouds, one after another, blossoming from the desert floor. These were the images that had terrified me in childhood when I saw them in black and white on television and in films at school—the flaming pillars of cloud that even today occasionally dominate the landscape of my nightmares.

The days of the duck-and-cover drills were over by the time I reached elementary school. But the fear the Bomb inspired remained potent, and I learned to dread clouds if they blew and stretched into mushroom shapes in the New Jersey sky. My heart stopped momentarily if a siren in town bellowed at any hour other than high noon. I squirmed uneasily when tests of the Emergency Broadcast System interrupted the cartoons I was watching on TV, and this sound still unnerves me, especially since the terrorist attacks of September 2001. My parents believed in the domino theory of communism as fervently as they believed in the infallibility of the Pope. One morning, when I was a freshman in high school, I walked to my bus stop wondering how many days I had left. My parents had reported over their poached eggs and mugs of Sanka that Russian ships had been spotted off the coast of—*where*? I can't recall the location. What I remember is the fear. Contained within the small frame of the video screen, Howie's mushroom clouds—"Isn't that one a beauty?" he marveled—looked like sea creatures drifting in an aquarium.

As we passed Nellis Air Force Base, Howie directed our attention eastward. "They're testing the new Predator aircraft there," he announced. "Maybe we'll see them." *Predator.* I cringed at the violence embedded in the word, at the brutality rooted in so much of the rhetoric of the Test Site. The *penetrator* bomb. The *ramjet* engine. But then I reasoned that, in an organization that assigns missiles names like *Minuteman, Patriot,* and *Peacekeeper,* at least the military is calling the Predator what it really is. We could not see the aircraft at Nellis. We trundled on toward the Test Site, where we paused for another security check at Mercury—a town with a toxic name, christened thus for a nearby mine and founded by the Atomic Energy Commission in 1953. After we were cleared, our bus lumbered past Mercury's facilities, which include a hospital, housing complex, movie theater, and bowling alley. Mercury, Howie told us, was where "the Russians"

stayed in the early 1990s, when they visited Nevada for a treaty verifi-
cation check.

As I listened to Howie, I noticed a particular cant to his language.
Russian scientists and technicians were called *the Russians*, a label that
fell out of common parlance only after the collapse of the Soviet Union.
"We broke 'em," Howie explained of the Russians and the 1996 Test
Ban Treaty. "We outspent the bastards." Bombs, I understood, are
devices that are activated by *mechanisms*. Explosions are *events*. And in
true utilitarian spirit, earth is *real estate*.

We drove through Frenchman Flat, an outdoor archive of the earliest events activated at the Test Site. It is a surreal landscape, a desert
dotted with structures such as a steel bridge—with one side bowed in
from the force of a blast—leading to nowhere. The Test Site was home
to a thirty-six-acre farm from 1964 until 1981, and it was here, Howie
informed us, that experiments were conducted on pigs, sheep, goats, and
other animals to see how exposure to radiation would affect them.[31]

"Pigs were the most common animal used in the tests because of
the similarity of pig skin to that of humans," I would read later in *The
Nevada Test Site: A Guide to America's Nuclear Proving Ground*. "The
pigs were dressed in various garments, including flight jackets, to test
materials for their radiation protection."[32] The image unnerves me
because it reminds me of George Orwell's *Animal Farm*, in which the
pigs, the most intelligent creatures in the barnyard, lead a revolution
against their farm's human owners. Toward the end of the novel the
pigs are walking on two legs and wearing clothing, and at the story's
conclusion they are brawling with humans and it's impossible to dis-
tinguish swine from men. I'm tempted to say that we have become a
nation of pigs—the United States is, after all, the most well-fed nation
and the one that consumes the most energy per capita—but the comparison seems somehow slanderous to the pigs, especially to those
sacrificed to Trinity's fearsome successors.

After Howie escorted us past the ghost farm, we stopped at Sedan
Crater, which is now on the National Register of Historic Places. Sedan
was detonated in 1962. This was what the DOE calls a Plowshares
project—a test for purportedly peaceful deployments of the Bomb for,
say, the construction of canals or harbors.[33] We stood on a platform
above the crater. "We wanted to see how much real estate we could

move," Howie explained. The event moved 12 million tons of it. "The 104 kiloton device was buried only 635 feet below the surface," reports the Center for Land Use Interpretation. "The explosion . . . created a crater which is 1,280 feet wide and 320 feet deep."[34] I had watched this detonation on Howie's video earlier that day; on the small screen, the cloud of earth blossomed like the marshmallows I toasted to burnt sugar on a stick at Girl Scout barbecues.

Nuclear Families and Hometown Big Bangs

Our bus growled over the Test Site's gravel roads, and Howie passed around binders crammed with DOE photos. There were the sea-creature mushroom clouds again, hovering over the desert. There were photos of soldiers sent into the Test Site's trenches to determine how they would operate during a nuclear attack.[35] But the most shocking photos were those of what I call the Nuclear family—life-sized mannequins that look like Barbie, Ken, and their son, Howdy Doody. They were nuclear war versions of crash test dummies, set up in prototype towns and bombed so that the bombers could estimate the effects of nuclear explosions on "a typical American community."[36] The photos tell their own story of 1950s culture and of who was really calling the shots. In one picture, a pony-tailed Barbie reaches toward a shelf of canned goods in a grocery store erected on the Test Site; clad in a halter top and miniskirt, midriff bare, she is as impossibly perfect as the Barbie dolls I played with as a child, a version of the dolls my little nieces play with today. Dressed like some Atomic Energy Commission guy's version of the ideal housewife, she is bombed. Another photo shows the mannequin family in flannel bathrobes, cheerfully cozying up to one another in a fallout shelter. In the most disturbing shot, Barbie, her clothes trimmed in white eyelet lace, and Ken are slumped over their dinner table, post-explosion, still grinning resolutely even though chunks of Ken's face have been chipped away. The Nuclear family lived in a village constructed on the Test Site; we drove by what remains of the town, pausing at the Nuclear family's house. It's a colonial, and the wood is bare because the heat of the blast, Howie told us, had blistered the paint from the exterior.

A few months before the Sedan event, the Atomic Energy Commission had detonated a smaller bomb called Passaic, which is the name of the river that threads through New Providence, the New Jersey town where I grew up. My parents moved our family there at about the time of these explosions. The town's two main streets grew along the lines of its churches. The Presbyterian church, at the intersection of the two roads, sprouted first, back in the 1600s when the place was called Turkey. According to local lore, the village was renamed New Providence after the beams of the church collapsed without injuring any of the faithful. Crooked brown headstones leaned wearily in the churchyard cemetery. A quarter of a mile up the road, the Catholic church fattened into an imposing, 1960s-style arena of stained glass and desert-tinted stone. The Lutheran enclave was stationed half a mile uphill near Bell Labs, an immense research complex where, in 1963, scientists picked up the microwave echoes of the Big Bang, which, as the poet Bob Hicok writes, was "the fucking / that got everything under way."[37]

Nineteen sixty-three was also a busy year for Nevada; the Atomic Energy Commission exploded forty-nine nuclear bombs, most of them at the Nevada Test Site, with a handful detonated at Nellis Air Force Base and one, the twelve-kiloton "Shoal," on military property in the agricultural community of Fallon. The DOE's records indicate that fourteen of these tests released radioactivity into the air, several of them off-site; of the fourteen releases, ten are listed as "accidental."[38] Does that mean that four of the releases of radioactivity into the air were deliberate? We had not then, as we have not now, figured out how to handle all of the nasty items in God's toolbox. And after Oppenheimer's Trinity, the phrase "Big Bang" could easily have taken on an antithetical meaning—the fucking that could shut the world down.

According to the Center for Land Use Interpretation, "radioactivity was released into the atmosphere and detected" outside the Nevada Test Site for at least twenty-nine of the underground tests. Radiation from the 1968 Schooner test was picked up across the continent in Montreal five days after the explosion. Schooner left a crater so large that NASA used the Test Site's pocked landscape as a moon-school to train the astronauts of Apollo 14, 16, and 17.[39]

Suburban Pastoral

Our lunar explorations barely registered on my radar; as a kid in New Jersey, I was more preoccupied with survival in my local landscape. New Providence was a small, working-class community. A handful of elegant houses stood like ancient relics in a town full of postwar residential developments—mostly ranch houses, split levels, and Cape Cods, with a few modest colonials. Sixteen miles west of Newark, where racial riots erupted during the civil rights movement and where Thomas Moran had created some of his great Manifest Destiny paintings of the Far West, our town straddled a belt of development bordering the edge of "the country." Across from the Catholic church, a meadow ran wild on the main street, and in springtime cornflowers stippled that meadow as if a chunk of sky had shattered in the grass. Cornflowers bloomed out of cracks in the sidewalks, and a blue ruffle lined the curbs of the valley. One of our neighbors raised ducks; another kept a hen and a crotchety rooster. Down the street, one family housed a flock of homing pigeons in a dovecote; evenings we watched them circling over the neighborhood. It was a one-horse town, with a speckled mare grazing in the field of the town's single working farm just a few blocks from our house. Some of our playmates' fathers worked for Bell Labs, the presence of which, our friends assured us, guaranteed that New Providence would be a top target in any nuclear attack.

Our parents sent their six children to the ironically named Our Lady of Peace Catholic School, where signs posted the way to the basement fallout shelter and the nuns prophesied our eternal damnation. A butchered Jesus hung on the front wall of every classroom. At home, a slaughtered Jesus dangled from his crucifix in our front hall and a Sacred Heart with a nimbus of blond hair kept mournful watch from the wall above our parents' bed. Our television bellowed the gunshots and war cries of cowboys and Indians, World War II movies, and the endless bad news of Vietnam, where more than fifty-eight thousand American service personnel and between one and two million Vietnamese would die by the time our newspapers carried photos of locals clinging to our departing helicopters, and where we were fighting because, like our parents, most of the powers in Washington believed in the domino

theory of communism.[40] The global menace of MAD loomed above
our kickball games as we awaited the sneak attack of the Soviet Union.

A History of Unnatural Destruction

The Test Site tour concluded at the Atomic Testing Museum, where
visitors' footsteps set off films of the Trinity explosion at Alamogordo,
an image that the environmental writer Rebecca Solnit describes as
"the postmodern sublime . . . the aesthetic of vastness, magnificence,
power, and fear."[41] Throughout the museum, I found the aesthetic of a
most uncivil religion—what we might call the *military sublime*. Think
of it as the technological sublime—the power of engineering—wedded with belief in a divinely ordained righteousness and framed in the
rhetoric of patriotism.

Here I learned that Nellis Air Force Base is named after a local hero.
The museum describes Lieutenant William Nellis as an "intensely
patriotic and exceptionally courageous pilot . . . [who] died scoring
hits on his target during the Battle of the Bulge in Belgium during
WWII." Since "scoring" is a word we also associate with sports and
sex, its use in this context of military heroism troubled but did not sur-
prise me. The feminist critic Annette Kolodny contends that Euro-
American men have dominated the landscape in the same ways that
they have dominated women, and that they have often used the same
language to describe both experiences of conquest.[42] Such language,
conjoining sex, sports, and war, erases the bodies of the dead and represents the battleground as a kind of gridiron. It also betrays the sort
of cultural forgetting against which the Gulf War veteran Anthony
Swofford warns in his memoir, *Jarhead:* "The warrior becomes the
hero, and the society celebrates the death and destruction of war, two
things the warrior never celebrates."[43]

The museum houses a collection of artifacts from the glory days of
atomic weapons testing. There are instruments, newspaper clippings,
maps, mannequin cousins of the Nuclear family. Early films about atomic
science portray atomic energy as a "nuclear genie," a brawny giant with
a public service job: warrior, engineer, healer, researcher. In a theater
that resembles a concrete bunker, filmgoers can imagine themselves as

reporters at News Nob, the Test Site's press observation station. At zero time, as the mushroom cloud bursts across the screen, the audience is hit with a shockwave of air and a vibe that zaps the legs and loins. I forced myself to remain even though I wanted to shoot from the room just as I had done months earlier in a tiny Montana museum, where I refused to sit in the dark and listen to a re-creation of the U.S. Cavalry shooting up the Nez Perce in the battle that finally provoked Chief Joseph's surrender. After the detonation, the Atomic Testing Museum's film features testimonials by former Test Site workers who describe themselves as "foot soldiers on the battlefield of the Cold War."

Throughout the museum, in films and stills, bloated mushroom clouds flash and roil in the sky. Some months earlier, I had stood in a San Francisco art gallery, transfixed before Chester Arnold's 2004 painting *A Natural History of Destruction*. It was part of a series that Arnold called "Reconstruction," in which several canvases depicted workers rebuilding the cities they had recently shattered. In *A Natural History of Destruction*, a beautiful city burns beneath a sky dominated by the violet form of an immense human brain. "The Brain – is wider than the sky . . . // The Brain is just the weight of God," Emily Dickinson wrote in 1863, one hundred years before the Limited Test Ban Treaty would force the mushroom clouds underground.[44]

Tucked into a corner is a modest display of Native American artifacts assembled by the Consolidated Group of Tribes and Organizations, seventeen branches of Paiute and Shoshone peoples who claim the Test Site territory as ancestral land. "This has always been our home," the exhibit reads. The classic photo of Rainier Mesa, with its hidden web of irradiated tunnels, hangs nearby.

The museum even has a touch of the Jeffersonian pastoral, with a silo from the Test Site's farm reconstructed as a theater, where I learned about the counterterrorism training operations currently conducted at the Test Site. And encased in glass is a fragment of a column from New York City's World Trade Center, with a placard reminding us of the Test Site's counterterrorism mission—a chapter of the nuclear narrative that struck this former New Yorker as particularly manipulative. Our nation first opened the nuclear tabernacle, after all. An assembly of powerful people had fostered a culture of fear, and I knew they wanted to keep me afraid.

The postgame party tone of the Atomic Testing Museum is neither surprising nor dated. Consider the tourists of 1950s Las Vegas, who gathered to watch the mushroom clouds blooming over the desert, marveling at the power so many believed we had harnessed for the good of humankind. Then consider the American-led invasion of Iraq in March 2003. At first, the big three television networks carried the story for hours, and their embedded reporters beamed home nighttime footage of our tanks slouching toward Baghdad. My television screen glowed green with the cameras' infrared light as the reporters marveled at the latest war technology, like tanks equipped with their own bridges to help them roll across crevassed desert terrain. Like the nineteenth century's Manifest Destiny paintings that glorified conquest of the continent, news reports in those earliest days broadcast images of the military sublime, American war machines encasing a force that I once imagined belonged only to God. The pictures of bodies and wrecked cities and villages would come later. Steven Newcomb, co-director of the Indigenous Law Institute, finds a disturbing parallel between our times and the nineteenth century. In 2004, writing about a 2002 speech on homeland security that President George W. Bush delivered from Mount Rushmore National Memorial, Newcomb remarked that Mount Rushmore "symbolizes the United States' effort to 'bring democracy' to the Plains Indian nations by bullets and warfare in the same way that the United States is now claiming to 'bring democracy' to Iraq by warfare . . . [and] to 'pacify' the Iraqi people."[45]

At the conclusion of our outing, Howie was beaming. President George W. Bush, Howie said, might reactivate the nuclear testing program, and the DOE might want "the experienced old-timers" to come back. Howie, who was about to embark on a cross-country drive with his wife of many years, had just signed the papers saying he would return to work if called. It was a troubling moment. Howie was a kindly man, about the same age as my father, and all day he had been speaking my father's cold war rhetoric. Howie was happy that his country still needed him. And why shouldn't he be happy? He came of age in the era of the Company Man and the long marriage. It must feel good to belong and to believe. But "the things that cannot be sustained are clear," writes David W. Orr in *The Last Refuge: Patriotism, Politics, and the Environment in an Age of Terror.* "The ongo-

ing militarization of the planet along with the greed and hatred that
feeds it are not sustainable. Sooner or later a roll of the dice will come
up Armageddon whether in the Indian subcontinent, in the Middle
East, or by an accidental nuclear weapons launch or acts of a rogue
state or terrorists."[46] I could have felt happy for Howie if he had been
talking about any other job.

Outside the borders of the Nevada Test Site and beyond the parking lot of the Atomic Testing Museum, does anyone really long for
the good old days of the cold war? Yet violence is woven into our
cultural mythology. Nevada's state slogan is "Battle Born," a phrase
that binds territory with bloodshed and liberty. Today, someone who
gets angry "goes ballistic." "We know the story of civilization; it can
be understood as a history of conquest, law-bringing and violence,"
writes William Kittredge. "We need a new story in which we learn
to value intimacy. Somebody should give us a history of compassion,
which would become a history of forgiveness and caretaking."[47]

On my way home from Las Vegas, I drove past Walker Lake, where,
during the late nineteenth century, the Paiute mystic Wovoka initiated
the ghost dance, a religious rite that envisioned a restoration of Indians
and game to their ancestral lands. The practice, adopted by tribes
throughout the West, was outlawed by the U.S. government, and it was
after ghost dancing that a band of Lakota were massacred at Wounded
Knee just a few days after Christmas in 1890. The Nevada Division of
Wildlife reports that Walker Lake is dying. As Nevada's 1894 boosters
envisioned, much of Walker River's flow has been diverted for agricultural operations. Walker Lake is "dropping four feet a year," resulting
in chemical changes that render it far less hospitable to the fish, waterfowl, and human communities it has supported for millennia.[48]

I stopped for gas and caffeine in the agricultural town of Fallon, a
community infamous for the leukemia cluster that has fingered seventeen of its children since 1997. The cause of the cluster is uncertain,
but researchers believe that it is at least partially attributable to environmental contamination.[49] In the restroom of the Fallon 76 Station,
I found a condom vending machine with a cheery sales pitch: "The
French"—the word "French" was blasted apart—"Tickler. Tickle her
fancy with the real thing. It's the patriotic thing to do!" About as patriotic as eating freedom fries. The machine was made in Korea, which,

according to the Korea Institute of Military History, still suffers from the aftereffects of the Korean War of 1950 to 1953, a conflict that killed more than 3 million North and South Korean soldiers and left nearly 2.5 million civilians "killed, injured, or missing."[50] A total of 54,246 American service men and women also died fighting the communist invasion of the North Koreans, backed by the Soviet Union, into South Korea.[51] During the Korean War, we managed to refrain from dropping nuclear bombs.

Flash Forward

It is tempting to dismiss the Test Site tours and the Atomic Testing Museum as memorials to a paranoid era. So much of Las Vegas is neon kitsch, but amid all the party favors, the Atomic Testing Museum is a political tool glorifying the beliefs that produced the bombscape on the city's outskirts. That ideology is neither innocent nor dead. On the fourth anniversary of the terrorist attacks on New York City and Washington, D.C., I would read in my local newspaper that the Pentagon was revising the "Doctrine for Joint Nuclear Operations." The new language of the nuclear doctrine would empower our military, with presidential approval, to use nuclear weapons in preemptive strikes.[52] Would Nevada's desert once again "bloom with atoms"?

Our recent record on the intelligence used to justify preemptive strikes is—in the most optimistic term I can apply here—faulty, and the Pentagon scuttled this controversial amendment in early 2006. Yet in 2005 and 2006, Nevada and Utah wrangled with the Department of Defense over the proposed Test Site detonation of Divine Strake, a non-nuclear weapon that scientists liken to a low-yield nuclear bunker buster that could be used in, say, Iran.[53] Las Vegas has attracted millions of tourists with the slogan, "What happens here, stays here," but the Test Site's downwind neighbors know that dust doesn't acknowledge the bombing ground's boundaries. Many feared the Divine Strake test would kick up irradiated dust from the decades of nuclear bombing. Some of us feared—and continue to fear—yet more war. "In place of a love for the historical rights and responsibilities of the nation, instead of creating community through inclusive and democratic measures," writes George Lipsitz, patriotism now "emphasize[s] public spectacles of power and private celebrations of success. It does not treat war as a regrettable last resort when all other means have failed, but rather as an important, frequent, and seemingly casual instrument of policy offering opportunities to display national purpose and resolve."[54]

Following more than a year of public outcry and bad publicity, the Department of Defense in February 2007 canceled the Test Site detonation of Divine Strake. But did the project really die? "Pentagon officials seemed entirely too willing to just give up on a test that, until [late February], they said was absolutely, positively necessary," one anonymous contributor observed in the *Las Vegas Sun*. The Department of Defense had admitted that the "test [of Divine Strake] could lead to the development of nuclear or conventional weapons." In October 2006, in the midst of the dustup, the Department of Defense revealed in a statement that it was "assessing other possible sites for the experiment."[55] Perhaps the test will be conducted elsewhere and under another name untainted by the public backlash that Divine Strake provoked.

And what of the name, Divine Strake? Since the mid-1970s, the Department of Defense has used a computer program to assign each bomb, "in theory, a discreet, descriptive, alliterative, non-insulting name." Even so, *Divine Strake* is a label from the lexicon of the military sublime, and it appears to have come from the same body of rhetoric that produced *Operation Infinite Justice* for the post-9/11 invasion of Afghanistan. Since many believe that meting out "infinite justice" is strictly God's job, the wordsmiths, in a collective "oops," hastily changed the name of that invasion to *Operation Enduring Freedom*. Divine Strake—the name and the bomb—emerged from the same zeitgeist that produced a president who claimed to have "consulted a higher authority," presumably God, before leading our nation into Iraq in a war framed in the language of American wars since the Revolution: *Operation Iraqi Freedom*.[56]

"The willingness to use nuclear weapons first requires giving them moral legitimacy," writes the Global Security Institute's Jonathan Granoff. Doctrines such as Mutually Assured Destruction "were designed to increase stability during the Cold War," Granoff observes, and "these ideas . . . give the veneer of moral propriety to the weapons."[57] My years as a cold war kid taught me to be an apostate and an iconoclast. The United States may have a diverse population, but our media have long reflected a government and society that prefer to portray our nation as Christian (recall how much attention the media devoted to Barack Obama's religious affiliation before the 2008 presidential election) and war as somehow holy. But the Gulf War veteran Anthony Swofford says that religion and the military are "incompatible." Swofford adds that "the high number of fiercely religious military people . . . are missing something. They're forgetting the mission of the military: to extinguish the lives and livelihood of other humans."[58]

What would a test of Divine Strake or any such a weapon accomplish? "Reliance on nuclear weapons for [deterrence] is becoming increasingly hazardous and decreasingly effective," write former secretaries of state George Shultz and Henry Kissinger, former secretary of defense William Perry, and former chairman of the Senate Armed Services Committee Sam Nunn. Neither will "a deterrent strategy" work against terrorist groups, they add. "Unless urgent new actions are taken," these security experts warn, "the U.S. soon will be compelled to enter a new nuclear era that will be more precarious, psychologically disorienting, and economically even more costly than was Cold War deterrence." Shultz, Perry, Kissinger, and Nunn "endorse setting the goal of a world free of nuclear weapons and working energetically on the actions required to achieve that goal." Their recommendations include "continuing to reduce substantially the size of nuclear forces in all states that possess them."[59]

In April 2009, President Barack Obama called nuclear weapons "the most dangerous legacy of the Cold War" and affirmed "America's commitment to seek the peace and security of a world without nuclear weapons."[60] In September 2009, when the United Nations hosted the Security Council Summit on Nuclear Non-Proliferation and Nuclear Disarmament, Shultz, Perry, Kissinger, and Nunn announced that they supported "the leadership of the U.S. administration in this effort."[61] At the summit, President Obama called "the spread and use of nuclear weapons" a "fundamental threat to the security of all peoples and all nations." He added that "nations with nuclear weapons have the responsibility to move toward disarmament."[62] In a promising move, the United States and Russia agreed in March 2010 to reduce, by roughly one-third, the number of "nuclear weapons that [both nations] will deploy."[63] President Obama had earlier conceded that the goal of "a world without nuclear weapons" would "not be reached quickly," and, indeed, the reduction still leaves thousands of nuclear weapons cocked.[64]

Paper Cranes

Imagine a trinity of the world's most powerful men fingering the atomic energy of the universe. Why would Truman, Churchill, and Stalin back the U.S. bombing of Japanese cities crowded with civilians? The imprisonment of more than one hundred thousand Issei and Nisei at concentration camps throughout the United States following the bombing of Pearl Harbor now seems like an attempt at ethnic cleansing without the outright genocide.

Manzanar National Historic Place, one former concentration camp, is just under two hundred miles from the Nevada Test Site, over the California border, and smack up against the jawbone of the Sierra Nevada. I traveled there alone in early autumn. This place once confined ten thousand children, women, and men of Japanese ancestry behind barbed wire. Prisoners lived in uninsulated barracks—a hardship in country that, at four thousand feet in elevation, means skin-scorching summers and bone-splintering winters. This is land that once belonged to the Owens Valley Paiute, who are also among the seventeen tribes laying claim to the region of the Nevada Test Site and who were removed from the valley by the U.S. military in 1863. Unlike the modest exhibit on Native American culture that stands like a footnote toward the back of the Atomic Testing Museum, Manzanar's former gymnasium now houses a sizable exhibit on the region's Native American history just inside its entrance. It also offers an unflinching look at our nation's history of violations of the civil rights of both minorities and anti-war protesters. "Where's the Guantánamo exhibit?" I asked the rangers at the reception desk. "A lot of people ask that question," a ranger answered. "Or they leave here saying, 'This can never happen again.'"

I prowled the grounds, poking among the ruins, amazed at the courage of these people who lived so bravely under such humiliating conditions. They built elaborate Japanese gardens throughout the camp.

And like patriotic Americans all over the country, Manzanar's prisoners planted two victory gardens—wartime community gardens that supplemented their food supplies, "ease[d] the strain on the nation's food production system," and served as symbols of loyalty to the United States.[65] The grounds are lush with wildflowers in spring, particularly around the old latrines, a ranger told me later. I found one trumpeted blossom and crouched to look at it. Broader than the palm of my hand, it was ragged and nibbled at the edges. Later, when I studied my map, I would realize that I had been kneeling in the ruins of one of the victory gardens.

I walked through a ghost garden, past empty ponds and wells, wondering how lovers managed in a place with so little privacy. Wind pulled my hair and yanked at my clothes and a sun dog rose above the Sierra Nevada. Beside a concrete pool, empty but for a puddle of water, stood a tree to which someone had tied a thick cord strung with paper cranes—a whole flock of pink, purple, and pale green paper birds that fluttered like aspen leaves in the wind. This is what we do in such places, most all we can do—leave pieces of pretty paper, bits of metal or cloth, portions of tobacco, as offerings.

In a grove of cottonwoods, I fingered the carcass of a tree that lay beside the road, bark worn away, exposed wood cracked and whorled as desert-baked skin. A pile of rocks rested against the roots of another cottonwood, and several feet above eye level, the bark bore the horizontal scars of a barbed wire fence. This was not the prison boundary, so these lines were not from the fence that had contained the prisoners. The rangers later told me that these markings may have been from a prisoner's garden or from the days when this camp was ranchland. I stood for several moments, stunned by these decades-old tracks of barbed wire on the body of a tree. Cold wind, fragrant with snow, blew down off the Sierra Nevada, reminding me of the ranch where I had lived and worked for only a few months, on the border of Montana's Teton National Forest at the front range of the Rocky Mountains. At the edge of that ranch was a cemetery that belonged to the Métis, a mixed-blood people who, during the late nineteenth century, were forced down out of Canada and over the border into Montana, where they were unwelcome refugees. The same old story, more or less, with a different set of characters.

But at Manzanar, I felt flushed with hope, because as far as I could tell, at least here we were not shying away from truth, not spinning it into some techno-teleological grand narrative of what passes for American progress. The aesthetics of patriotism celebrated at the Nevada Test Site and the Atomic Testing Museum encourage a sense of community motivated more by fear than by love of country. They endorse a military sublime informed by frontier attitudes of desert as wasteland and twentieth- and twenty-first-century visions of Nevada as nuclear space. They conjure up an Old West scenario for the nuclear age, with the clock set at one minute to high noon and the good sheriff at the ready, holster packed with MX missiles. At Manzanar, the aesthetics of patriotism are garden ruins, ropes of paper cranes, and a renovated gymnasium housing the complicated stories of a national community and of the peoples who once lived here.

Snowmelt had engraved the outline of a tree across the face of one of the mountains of the Alabama Hills. Myopia flattened the carvings into the blueprints of the earth: the patterns of fern leaves, fish bones, fungi, mineral veins, riverine circulatory systems, trees. As I drove away from Manzanar, I thought about how the surface of the earth remembers water and wind and nuclear bombs. How the events of our lives etch maps onto our brains and bodies. How the ghost of a barbed wire fence graces the bark of a cottonwood. How our bodies harbor contaminants that imprint their signatures onto our cells. How cities and islands remember blast and burn.

Home

My friends in Reno cringed whenever I mentioned the Test Site. "Why did you want to go *there*?"

I had hoped to make some sense of the technological barbarity that was spawned of fear and a lust for power but labeled in the name of peace. I had hoped to gain some understanding of the terror and violence of the cold war era, when the name of the Bomb replaced the name of God. But none of it makes any sense, and our current times are, on some levels, even more violent than those I grew up in. Nuclear weapons remain "at the center of the American dilemma," observes James Carroll in *House of War*, a study of the Pentagon from the cold war to the present.[66] Of our nation's early twenty-first-century military policies, Carroll cautions, "Beware the House of War when understood as the House of God."[67]

At Reno's Nevada Museum of Art, there is a gallery I often visited that is entirely filled with the painter Robert Beckmann's eight-canvas series, *The Body of a House*. Beckmann, who grew up in Las Vegas in the 1950s, based his paintings on stills from a Test Site film of the disintegration of a house taken when "Annie," a sixteen-kiloton bomb, was detonated on St. Patrick's Day in 1953. The Atomic Energy Commission wanted to see how such a home would weather a nuclear assault, and perhaps this house is as much the "House of War" as James Carroll's Pentagon. Like those still standing on the Test Site, this house is a colonial, which leads me to wonder who chose the design of the structures erected for Test Site nuking, and why. There is an onanistic poetry to the images of this iconic American home—a *colonial* erected on a landscape for a technological experiment that is the self-annihilating climax of Manifest Destiny—collapsing beneath the force of the Bomb. In the first painting, the facade of the house is illuminated by the brilliance of the blast. But for its desert surroundings, the house looks very much like the one I grew up in, and I would walk around the gallery observing my childhood home, from canvas to canvas, progressively burn, melt, disintegrate, and blow away. What remains is war—three letters, a trinity, an atom.

3
Timber Culture
Scenic Oregon and the Aesthetics of Clearcuts

Everything rises
from mere roots
anchored in the brown
bone of the earth.

Robert Davis, "Into the Forest"

I keep a postcard of Oregon taped to the wall above my desk. The image is classic Americana: In the distance, bright sunlight illuminates Mount Hood's snow-mantled shoulders. Richly timbered mountain slopes give way to green and gold farmland, where a neat fence frames the yard around a red, tin-roofed barn. Beyond an orderly bristle of orchard, a farmhouse nestles in a shelterbelt. A ruffle of cattails frames the foreground. Behind the mountains, as if typeset by the hand of God, the word OREGON rises from the horizon to the roof of the sky. The scene looks rich and inviting, but the postcard reveals a human-imposed geometry. The timberlands beyond the farm have been clearcut over and over again; the entire range is a patchwork of various stages of regrowth. Arranged in the manner of the Hudson River school painters, the postcard is a postmodern vision of a pioneer's dream, cramming American sublime, pastoral, and utilitarianism onto 4 x 6 inches of paper that might have originated in Oregon's trees.

Asphalt Patriotism

My postcard reflects the irreconcilable tension between views of landscape as both aesthetic and material resources. I took a closer look at the postcard landscape on a recent drive, and after Nevada's raw-boned grace, western Oregon seemed an extravagance of green. I was guzzling scenery, a pastime inherited from eighteenth-century British tourists who enjoyed carriage rides and walks in the country while seeking picturesque landscapes. Serious picturesque tourists carried a Claude glass—a convex mirror, generally tinted black or green—in which they regarded the view. They did not look directly at the landscape; rather, they turned their backs to it and looked at its reflection in the Claude glass, which condensed, framed, and tinted it into an idealized representation of what was actually on the ground.[1] They preferred aestheticized constructions of nature over the real thing, establishing a tradition still at work in the United States today, with artful representations of the American landscape—like my postcard or any number of calendars, magazines, webpages, or books containing photos of landscape scenery—forming the panorama of national mythology. That panorama is also etched in asphalt. Today our 150 National Scenic Byways are designated "based on one or more archaeological, cultural, historic, natural, recreational and scenic qualities."[2] I had business in Oregon, but I also planned to treat myself with a drive along one of our most famous scenic byways—Pacific Coast Highway 101.

The American pastime of scenic tourism predates the automobile, and it became hip during the nineteenth-century's nation-building era. The popular thousand-page *Picturesque America*, published in 1872, featured stunning engravings of scenic places across the nation. The text, writes Sue Rainey, "promoted and reinforced a resurgence of nationalism rooted in the homeland itself rather than in the institutions of democracy."[3] For post–Civil War Americans, an appreciation of the features of their nation's terrain helped to foster a sense of commonality that transcended lingering sectionalism.[4] The historian Marguerite Shaffer notes that *Picturesque America* "defined a tourist gaze that . . . reduced the nation to a series of framed views, an assemblage of objectified landscapes."[5]

Picturesque tourism in the United States was succeeded by what Shaffer calls "national tourism," a trend toward carefully constructed "geographical consumption" of places of ideological significance. "In teaching tourists what to see and how to see it, promoters invented and mapped an idealized American history and tradition across the

American landscape, defining an organic nationalism that linked national identity to a shared territory and history," Shaffer writes, adding that "advocates promoted tourism as a patriotic duty."[6] Much as nineteenth-century American landscape paintings and prints had educated their viewers about landscape aesthetics and their ancillary ideologies, "national tourism" instructed travelers in the relationship between their landscape and the dominant American story. And just as expansionist-era American landscape paintings represented sublime landscapes "contained" by baroque frames and implicitly mastered by civilization, patriotic tourism represented designated places through the ideological frame of national mythology.[7] Shaffer remarks that "automobile touring rested on a romantic theory of nationalism" that "linked the nation to land and soil."[8] It was a nineteenth-century idea updated for evolving technology, with highways connecting storied places into a coherent saga.

By the mid-1960s, the Department of Commerce identified "driving for pleasure [as] the nation's most important outdoor recreation activity" and began pushing for a national scenic highway program.[9] One "objective of such an effort is to clarify and strengthen the motorist's image of the environment, to offer him a picture which is well-structured, distinct, and as far reaching as possible," the Department of Commerce proposed, revealing how deeply nineteenth-century landscape aesthetics remained embedded in mid-twentieth-century American culture.[10] The Department of Commerce's proposals were not entirely aesthetic, however; they were also motivated by highway safety and cold war concerns. "Roads serve national defense," Secretary of Commerce John Connor argued in 1966. Scenic highways would enable the rapid evacuation of densely populated areas and would facilitate the movement of troops and equipment during a war or natural disaster. Secretary Connor, apparently unfazed by New Jersey's juxtaposition of the giant Pabst beer bottle and the thousands of graves of Holy Sepulchre Cemetery, lauded the Garden State Parkway for its beautiful landscaping and outstanding safety record. He also doled out high praise to the state of Oregon.[11]

One of the first states to recognize that its scenery should be protected, Oregon was decades ahead of the federal government in this effort, and began to align roads with spectacular scenery—and to

attempt to preserve that scenery—during the early decades of the
twentieth century.[12] "Governor Olcott, alarmed over logging along
the Seaside-Cannon Beach road on the north Oregon coast, asserted
in 1920 that it was the 'patriotic and civic duty' of Oregonians to 'preserve our wonderful natural surroundings,'" writes David Peterson
del Mar. "The public generally supported the governor, agreeing that
weekend drives through Oregon's countryside ought not to be spoiled
by clear-cuts."[13] Samuel Boardman, who became Oregon's parks
superintendent in the 1930s, is credited with establishing a strand of
state parks along Pacific Coast Highway 101, which is studded with
scenic turnouts that face the ocean. Boardman also fought for timber
fronts along roadside clearcuts.[14] Timber fronts provide a screen of
greenery—sometimes as transparent as old lace—between the road
and any logging on the adjoining turf.

Into the Old Growth

I intended to examine some of western Oregon's clearcut corners, and
my first stop was Eugene, where I joined about twenty other researchers on a summer afternoon and rode in a caravan to the Andrews
Experimental Forest, a 16,000-acre parcel of the 1.7 million-acre
Willamette National Forest on the western slopes of the Cascades.[15]
Like the Nevada Test Site, the Andrews Experimental Forest is a
massive outdoor laboratory, but its experiments are constructive
rather than destructive in nature—although those opposed to all logging would disagree with me. It is also one of twenty Long Term
Ecological Research Sites, a National Science Foundation–supported
network of ecosystem study areas. Fred Swanson, a Forest Service
research geologist who has studied and worked in the region for more
than thirty years, several as lead scientist at the Andrews, piloted our
van. Swanson first came to Oregon as an undergraduate researcher in
the 1960s and quickly realized he had found a rock hound's paradise.
The state, he said, had everything a budding geologist could want:
volcanoes, steep terrain, earthquakes, major floods, and landslides.
He was lucky, I thought. By fluke of a family decision made shortly
before I was born, I just missed growing up in Oregon, and I have

often wondered what we passed up. Maybe we kids would have learned to swim in rivers rather than chlorinated pools, living an outdoor life instead of one circumscribed by asphalt and lawns. It wouldn't have been much different in Oregon, my sister says, and I know she's right. It would have been a similar story with different trees.

Swanson drove east from Eugene through Springfield. As we cleared town, farms spread out across the flats. Conifers and deciduous trees mingled on the hillsides and the road followed the curves of the McKenzie River. "This looks a lot like New Jersey," I said.

Swanson, who grew up in the Mid-Atlantic region, was taken aback. "In a little while, you'll see something that looks nothing like New Jersey," he answered, and within the hour my fellow researchers and I were surrounded by silvery western hemlock, western red cedar, Pacific yew, and five-hundred-year-old Douglas fir trees sheathed in corrugated bark. Guided by Swanson and Kathleen Dean Moore, a professor of philosophy at Oregon State University and director of the Spring Creek Project, an organization of environmental scientists, philosophers, and writers, we were hiking Lookout Creek Trail in the Andrews. Swanson is well over six feet tall and accustomed to working among giant trees; straight-backed and in his sixties, he moved along the trail with the youthful agility and vigor typical of rangers past their greenest years. Moore is petite, a blonde Renoir whose blue gaze focuses with hawklike intensity on whatever she regards. She knows this forest intimately, and later I would see her smile blissfully as she nestled against a Douglas fir as if its mossy roots were the arms of a favorite chair.

Swanson said that the language we use to describe such places reflects evolving social conceptions of old growth—trees more than two hundred years old. Landscape ecologists now call forests "the most species-rich environments on the planet."[16] But in the 1960s, Swanson said, old-growth forests such as the one in which we stood were called "large saw timber" or "biological deserts." The trunks of trees living and dead were powdered with pale green lichen, and lichen trailed like knotted hair from the branches. The scent of smoke from a controlled burn tinged the forest air. The forest around us was dim; shafts of sunlight cut through gaps in the thick filter of forest canopy—what foresters call the overstory. The understory is the lowest layer of foliage, between the middle story and whatever covers the ground. My colleagues and I had come to the Andrews to observe a corner of an old-growth forest, to learn about some of the long-term experiments in progress at the Andrews, and to think about our complicated relationships with old-growth forests and logging. Our nation's early stocks of timber provided the plotline in a chapter of the fables of American progress, and much of what remains of our old forests endures by stay of execution.

Thomas Cole and the Question of Progress

When we think about logging, many of us think first about the Northwest. Fly over northern California, Oregon, Idaho, Montana, Washington, or southeast Alaska and you will see that those sublime mountains beneath your plane are harlequinned with clearcuts. For generations, forests in the Pacific Northwest fed the bulk of American demand for lumber, and controversies over logging in northwestern states have figured prominently in national media over the last two decades. The conservation biologist Peter J. Bryant writes that in "Northern California, Oregon, and Washington over 90% of the ancient trees are gone."[17] But Americans began logging in earnest in the Northwest in the late nineteenth century, after devouring much of the old-growth woods on the East Coast and hacking apart the forests of the Great Lakes region.[18]

Thomas Cole was horrified by the logging around his Catskills home. Those "copper-hearted barbarians are cutting all the trees down in the beautiful valley on which I have looked so often with a loving eye," Cole bristled in an 1836 letter. "If men were not blind and insensible to the beauty of nature[,] the great works necessary for the purpose of commerce might be carried out without destroying it."[19] In Cole's *View from Mount Holyoke, Northampton, Massachusetts, after a Thunderstorm—The Oxbow* (fig. 3), painted the same year the artist wrote these words, the sublime wilderness recedes before an incoming tide of development. The painting depicts a storm sweeping like a curtain across the canvas, unveiling a pastoral landscape of cultivated fields that border a forest of mixed hardwoods. A river flows across the canvas, curving into an oxbow or, as Angela Miller notes,

Figure 3: Thomas Cole, *View from Mount Holyoke, Northampton, Massachusetts, after a Thunderstorm—The Oxbow*, 1836. Oil on canvas, 51 ½ x 76 in. (130.8 x 193 cm). The Metropolitan Museum of Art, Gift of Mrs. Russell Sage, 1908 (08.228). Image © The Metropolitan Museum of Art /Art Resource, NY.

into the shape of a question mark. On the wooded face of a low mountain in the background, the terrain is marked with what Miller identifies as "plough markings" signifying the Hebrew letters for "the Almighty."[20] But given the angle of the slope, the placement of the clearings in forested terrain, and Cole's words in his letter, we might well imagine these markings as scars of the logging that so distressed the artist, and we might read the river's question-mark shape as an inquiry about the costs of development.

Cole's *Course of Empire* series, completed the same year as *The Oxbow*, pointedly questioned Jacksonian democracy and America's path toward progress. Cole "was appalled by the Jacksonian frontier spirit, built upon the commercialization of the American wilderness with steam engines, sawmills, and man-made canals," writes the art critic James Cooper, who notes that the artist "feared that empire building in the New World would lead to its moral and spiritual decline."[21] If—in accordance with the transcendentalist spirit of the times—the landscape was both the dwelling place of God and the manifestation of his beneficence and power, then what would happen to American character as that landscape was developed? Some of Cole's works appear to have addressed that question while raising questions of their own. Across five canvases, *The Course of Empire* presents a single location over five successive eras. We observe as its woodland human communities evolve from their hunter-gatherer lifeway to establish a pastoral republic that develops into a flourishing city, which Cole depicted in the gaudy glory of prosperity and, subsequently, under siege by an enemy nation. In *Desolation*, the final painting, the wilderness reclaims the city's ruins and the humans have vanished entirely. Like *The Oxbow*'s questioning river, Cole's *Course of Empire* paintings seem to ask us, "What are we doing? Where are we going? Where will all this development—what we call 'progress'— lead us?" Cole's contemporaries greeted *The Course of Empire* with enthusiasm but failed to imagine that history might repeat itself on American soil. The message of the series was lost on them.[22]

Cole understood that the trees in his beloved valley had been cleared to make way for the fields from which he ate, and he found those fields beautiful, if not wild. "Where the wolf roams, the plough shall glisten . . . [and] mighty deeds shall be done in the pathless wilderness," Cole wrote in his "Essay on American Scenery" in 1835. "Yet I cannot but express my sorrow that the beauty of such landscapes [is] quickly passing away," he added wistfully.[23] Even as he recognized the need for development, and even as he portrayed cultivation as beautiful, Cole remained disturbed about the encroachment of development on the wilderness. Pastoral may be appealing, even picturesque, but it is, at base, utilitarian, and Cole sensed that the wilder beauty of the American landscape was irrecoverable except on canvas.

An Empire Made of Wood

Cole's was not the sole voice crying at the edge of a vanishing wilderness, and the problems that timber consumption created were not exclusively aesthetic. Nineteenth-century Americans were constructing an empire made largely of wood, and scientists, landscape architects, and industrialists alike were alarmed by our nation's insatiable hunger for timber. In a report published by the Massachusetts legislature in 1846, George B. Emerson, a writer, educator, and relative of the transcendentalist philosopher, warned that his state had already depleted much of its forestlands and was axing up those of its neighbors. "Every year we are more dependant [*sic*] on Maine and New York, and some of the Southern States. . . . Even these foreign resources are fast failing us. Within the last quarter of a century the forests of Maine and New York, from which we draw our largest supplies, have disappeared more rapidly than those of Massachusetts ever did," he fretted. "In a quarter of a century more, at this rate the supply will be entirely cut off."[24]

The nation's gaze, seeking timber and other resources, had already shifted westward. In his 1836 essay, *Nature*, Ralph Waldo Emerson wrote that "nothing in nature is exhausted in its first use. . . . In God, every end is converted into a new means." Commodity alone, Emerson believed, "is mean and squalid. But it is to the mind an education in the doctrine of Use, namely, that a thing is good only so far as it serves."[25] Emerson was writing during the age of Jacksonian expansionism, and many of his contemporaries, and those who followed, seem to have embraced his enthusiastic utilitarianism without the philosopher's accompanying reverence for the source of those commodities. Horace Greeley, editor of the *New York Tribune*, traveled overland by rail in 1859 and published a series of newspaper columns cataloging his experiences. "In Greeley's view, land had no value if it could not be used for farming or for growing trees to form stout farmhouses or railroad ties," notes the historian Anne Farrar Hyde. Hyde adds that the editor fairly gloated when he reached the Sierra Nevada. "'Almost every road is covered by giant, glorious pines!' Greeley exulted. 'In short, I never saw anything like so much good timber in the course of any 75 miles travel as I saw crossing the Sierra Nevada.'"[26]

"In the timber-rich Pacific coast country stretching from California's Humboldt Bay north into British Columbia and the Alaskan panhandle," writes the environmental historian William G. Robbins, "the forest resource has always been the centerpiece of the area's economic

culture."[27] Greeley's assessment reflected expansionist-era economics. "Resource capitalists . . . saw west coast timber as an ideal field
for investment," the historian Richard Rajala tells us. "The westward
movement of timber capital began in the 1880s," he adds, "and accel-
erated in the next decade."[28] Walt Whitman's final edition of *Leaves of
Grass*, published in 1891, included "Song of the Redwood Tree." The
poem is an anthem of American expansionism in which the redwoods
sing of their consciousness and souls, but graciously surrender their
lives so that humankind, "a superber race," may build an empire from
their substantial bones.[29] In a nation hungry for timber, writes Charles
Wilkinson in *Crossing the Next Meridian: Land, Water, and the Future of
the West*, "the spruce, cedar, redwood, and Douglas fir stands [of the
Pacific Northwest] amounted to green gold."[30]

"The building [of the railroads], their ties and trestles, consumed
a vast amount of lumber," observes David Peterson del Mar. The railroads also changed the nature of Oregon's economy. "By the 1890s
Portland was sending a large fraction of its timber exports east, along
the Northern Pacific [Railroad], rather than west, by steamship,"
del Mar adds. "This shift to rails . . . helped boost Oregon's timber
economy at the turn of the nineteenth century. It also jump-started
eastern Oregon's timber industry."[31] Eastern timber had built great
stretches of the railroad lines into the West, and western lumber
flowed back out. Karen Arabas, an environmental scientist, and the
political scientist Joe Bowersox note that lumber companies logged
their western landholdings so aggressively during the early decades
of the twentieth century that "by the 1930s . . . their forests were
severely depleted," and much of the cutting had shifted to national
forests by the start of World War II. "As the nation returned to a
peacetime economy in the late 1940s, the federal government continued to expand its harvest rates on public lands to meet homebuilding
and paper needs," they observe. "By the 1950s, Oregon, Washington,
and Northern California accounted for over 50 percent of the forest
products produced in the nation. By the 1980s, nearly 10 billion board
feet of timber per year were coming off federal forest lands."[32]

The flood of western lumber slowed dramatically in 1994 with the
implementation of the Northwest Forest Plan (NWFP), organized by
presidential mandate after the northern spotted owl was listed as an

endangered species. Controversial since its inception, the plan was designed to balance the needs of old-growth ecosystems, local communities, and the timber industry. According to John Kitzhaber, a former governor of Oregon, one shortcoming of the NWFP is that it was developed without much public involvement. "Consequently, there is still a great deal of tension surrounding it," he writes. The plan, Kitzhaber says, "acknowledges that any additional harvest of old-growth forests will further threaten sensitive species dependent on these forests, [but] expect[s] this same old-growth forest to provide 80 percent of the projected timber volume over the next twenty-five years." The goals of the plan appear to be incompatible. "By trying to maintain parity between multiple goals, the NWFP falls short of adequately addressing any of them and is not politically sustainable over the long run," Kitzhaber concludes.[33]

The George W. Bush administration promoted the Healthy Forests Restoration Act with what it called "Healthy Forests: An Initiative for Wildfire Prevention and Stronger Communities." Signed into law in 2003, the act has sought to meet the needs of industry by addressing what the Bush White House called "forest health." "Healthy forests" sounds innocuous compared to "logging," which is essentially what the plan proposed. Oregon, like other western states, is struggling with the consequences of decades of fire suppression that have turned some forests into tinderboxes—an outcome, in part, of viewing forests primarily as timber assets that have not yet been liquidated. Yet overgrown timber is just one—albeit critical—factor in the severe western fires of the last several years. Scientists also point to global warming, which is causing longer and drier fire seasons.[34]

Elements of a Forest Life

At the Andrews, we hiked deeper into the forest surrounding Lookout Creek Trail. I picked up a fragment of Douglas fir bark and it crumbled like old paper in my hands. Further in, an immense Douglas fir had fallen across the trail, and a maintenance crew had cut a chunk out of the trunk to keep the path clear. We passed through. It smelled like a new house. The trunk had cracked through the heartwood, and the break split into the shape of a cross that radiated out from the center, slashing through the tree's rings and pointing in the four directions. "The moral law lies at the centre of nature and radiates to the circumference. It is the pith and marrow of every substance, every relation, and every process," Emerson wrote in *Nature*.[35] I know the shape of the break was determined by laws of physics and had nothing to do with the cultural symbols I found myself imposing on the cracked heartwood. But as I read over Emerson's words, I wonder if some imperatives transcending time and science were encoded in the ruptured rings of that tree.

When we reached a bend in the trail at the base of a sheltered hollow, Kathleen Dean Moore instructed all of us to find a comfortable spot, sit quietly, and pay attention to the forest. The group fanned out. Some rested on rocks or fallen boles or nested in the laplike roots of big trees; others reclined on beds of duff and pillowed their heads against mossy logs; others climbed the slopes and braced themselves against standing trees. I tried to open my senses to the forest, but unaccustomed to sitting still and not given to quiet reflection among large groups of people, I frittered about, restless. I picked up a flake of a Douglas fir cone and noted how its lobe-like markings resembled the figure of a moth, much as later, when I walked this trail alone, I would observe a moth, camouflaged as a scrap of tree bark, settle by my boots. I was grateful to be at the Andrews and in the company of these researchers, but as much as I wanted to know this forest, I could not quite surrender to Moore's instructions in this unfamiliar place.

I should have realized that I was only a few steps from home. Every night I bed down in a glen of forest products. My wooden cabinets are stocked with printer paper and file folders, and I deplete several packs of both every year. A wooden basket holds a continually shifting stash of magazines and newspapers. I pour milk and juice from cardboard containers and eat meals seated on a wooden chair pulled up to a wooden table. Books, boxed in cardboard, come in the mail nearly every month, and the walls of my tiny apartment are lined with particleboard bookcases that sag with an overload of books.

One of those books is *Elements of Forestry*, a popular college textbook that was published in 1914, when Americans were consuming approximately five hundred board feet per person per year, and just nine years after our public forestlands had been transferred to the Department of Agriculture.[36] The text, which went through several successive printings, reveals a great deal about cultural attitudes toward forests that prevailed throughout much of the twentieth century.[37] The authors, Frederick Franklin Moon and Nelson Courtlandt Brown, both professors of forestry at New York State College of Forestry at Syracuse (Moon would eventually become dean of the college, and Brown would serve three terms as acting dean), defined forestry as "the art of raising repeated crops of timber on soils unsuited to agriculture, and properly utilizing the products of the forest."[38] They added that "non-agricultural areas which would otherwise lie idle are thus made to yield a revenue that will pay taxes and interest on the capital which they represent." Timber was a "crop," soil was for growing crops, forests were to be "utilized," and land was a symbol for "capital." Decades after Horace Greeley's trip west, the professors were chanting the editor's land management slogan: farm it or grow trees. They believed that "forestry should be taught to every student of agriculture" enrolled in "state agricultural schools and colleges." "The growing of wood is the chief end of the forest, for without timber, railroads, cities, and in fact civilization would have been impossible," Moon and Brown advised, echoing the progress-of-empire chorus caroled by their fathers and grandfathers and Walt Whitman, and chronicled in so many nineteenth-century landscape paintings.[39]

According to Moon and Brown, "the other ends served by the forest are incidental for the most part."[40] The foresters dedicated their book to Theodore Roosevelt, and they wrote it during the post-frontier age of masculine anxiety. Perhaps fearing any sappy association with the transcendentalists of the previous century, whose meditations tracked the divine Over-soul in their local forests, or with the ecstasies of the preservationist John Muir, Moon and Brown insisted that "forestry is not based on sentiment nor upon the desire to preserve the forests for their beauty, but that the entire forestry movement is founded on sound economic principles."[41]

The professors' language demonstrates how deeply utilitarian models were embedded in constructions of the environment by the start of the First World War. They still are. Scientists and policy-makers now speak of environmental "assets" that provide us with

"ecosystem services" and "cultural services," balance sheet rhetoric that reflects a master-servant relationship between humans and
the environment, an ideology we can trace to the book of Genesis,
when God designated Adam master of the earth.[42] In a 2003 speech
promoting the Healthy Forests Initiative in Redmond, Oregon,
then-president George W. Bush called American forests one of our
"greatest assets."[43] During National Forest Products Week, the third
week of every October, he called our forests "a source of pride for all
Americans" and "an important symbol of our Nation's beauty and
economic strength"—language that reflects both the aesthetics we
revere and the utilitarian pressures we place on our landscape but
does not acknowledge the tension between them.[44]

Gifford Pinchot and the Aesthetics of Conservation

Gifford Pinchot, who served as the Forest Service's first chief from
1898 through 1910, tried to reconcile the tension between the utilitarian and the aesthetic.[45] The environmental historian Char Miller
notes that "Pinchot acted as the main publicist for what historians
call 'utilitarian conservationism,' the belief that natural resources
such as lumber, coal, and water should be sustainably used and that
the federal government should regulate use."[46]

Pinchot was born into a family that nurtured the fledgling field of
American forestry and treasured American landscape art, and the
aesthetic tastes his parents cultivated evidently shaped, in part, the
future forester's utilitarian philosophy. "Gifford Pinchot has long
been accused of practicing a kind of dry-goods forestry, a denuncia-
tion of the profession's supposed exclusive utilitarian origin and commercial orientation," argues Miller. "But Pinchot may well have built
his later studies in forestry's scientific language upon an already established aesthetic vocabulary; in his hands, the forester's tools could be
a paintbrush."[47] The future forester's father, who worked as a wallpa-
per merchant and manufacturer at a time when Americans were wall-
papering their homes with landscape scenery, served as chair of the
art committee of the Metropolitan Museum of Art, and would found

the Yale School of Forestry in 1900.[48] James and Mary Pinchot, Gifford's parents, befriended several of the Hudson River school painters, and a substantial collection of the artists' works graced the walls of their homes in Milford, Pennsylvania, and Washington, D.C. Gifford Pinchot was so steeped in nineteenth-century American landscape aesthetics that they were encoded in his name the day he was born; the future forester was called after Sanford Robinson Gifford, the Hudson River school painter.[49] Pinchot died in 1946. A few years after the 1960 death of his widow, Cornelia, the Forest Service was given the family's Pennsylvania estate, along with its collection of approximately two hundred American landscape paintings, all of which have since disappeared. In the early 1960s, Hudson River landscape paintings were considered "out of fashion," and the Pinchot family believes that the paintings were "dumped into a hole on the [Pennsylvania] property and buried."[50]

One painting that did not get dumped was Sanford Robinson Gifford's *Hunter Mountain, Twilight*. Painted in 1866, a year after Gifford Pinchot was born, it hung above the family's fireplace for several decades.[51] The painting combines American pastoral—cattle grazing in the middle ground along a shimmering stream, a rural farmhouse tucked into a hollow—with a golden sky above a range of mountains disappearing into the distance. The foreground of the painting is filled with stumps. The art historians Kevin Avery and Franklin Kelly observe that the region was known for its tanneries and that "the locale [of the painting] may have been a site on which hemlock trees were stripped for the tanning industry."[52] A stretch of forest has been cleared, perhaps to build the farmhouse. Scholars speculate about the meaning the artist intended for *Hunter Mountain, Twilight*. Some view the painting as a commentary "expressing the nation's current mood," indicating the "somber peace in the Union Victory."[53] Others see the "denuded foreground of the picture, and the fading light in which the scene appears . . . [as] complementary elements in an elegy for the untouched wilderness of an America fast disappearing."[54] Perhaps Sanford Robinson Gifford shared Thomas Cole's ambivalence about development. "The axe, a symbol of economic progress and cultural poverty, of conquest and death, was double-edged," observes Char Miller.[55] "Gifford's job," Miller says, "was to put the trees back."[56]

The painting's message is ambiguous. Perhaps the yellow sky of *Hunter Mountain, Twilight* indicates the twilight of a golden age of wilderness. Yet in his 1782 paean to the American pastoral, *Letters from an American Farmer*, Hector St. John de Crèvecoeur had written, "Here [one] beholds fair cities, substantial villages, extensive fields, an immense country filled with decent houses, good roads, orchards, meadows, and bridges, where an hundred years ago all was wild, woody, and uncultivated!"[57] Sanford Robinson Gifford's farmhouse is a cornerstone of the agrarian republic. Its windows aglow with firelight, the farmhouse looks warm and inviting. A single cruciform star glows in the golden sky, possibly indicating blessing and a teleological vision of the nation's recovery after the Civil War. The star is tiny, yet it gleams.

Because it hung in the Pinchot home for decades, *Hunter Mountain, Twilight* formed one of the focal points in the rich aesthetic backdrop of Gifford Pinchot's life. The images portrayed in the painting would surface in 1905 in a document that the Forest Service now refers to as the Pinchot Letter, penned on the day when the forest reserves, which had been under the rubric of the Department of Interior, were transferred to the Department of Agriculture, where they officially came under Pinchot's domain.[58] As Charles Wilkinson notes, Pinchot asserted that the forest reserves were for "productive use," and that "the continued prosperity of the agricultural, lumbering, mining, and livestock interests is dependent upon a permanent and accessible supply of water, wood, and forage." Pinchot stressed that these resources must be protected and "wisely used for the benefit of the homebuilder first of all, upon whom depends the best permanent use of lands and resources alike."[59] With his references to wood, water, forage, and the homebuilder, one imagines that Pinchot might have drafted this letter while sitting at home, facing *Hunter Mountain, Twilight*.

American Dream

Until the housing slump began in 2006, the homebuilder was still consuming a great portion of our lumber, and if "the axe was double-edged," as Char Miller writes, then so is the chainsaw. In 2005, we produced approximately 52.3 billion board feet of lumber.[60] The bulk of it—22.8 billion board feet—came out of the South. The West provided 20 billion board feet, the next largest share.[61] James L. Howard, a Forest Service economist, notes that per capita consumption of lumber in the United States in 2005 "was 255 [board feet], above the high of 253 bf set in 1987." Overall, we consumed about 75.6 billion board feet, and 60 percent of that went to housing. Howard reports that housing starts in 2005 "were 2,215,000 as sales of new houses set a new record in 2005 of 1,283,000 units." New housing construction, he adds, "accounts for more than a third of U.S. annual consumption of softwood lumber and structural panels and for consumption of substantial volumes of other softwood and hardwood products."[62] The Department of Housing and Urban Development calls homeownership "an important part of the American Dream," and American dreams are expanding.[63] The new houses are about 97 percent softwood, an increase over previous years that Howard attributes to "a decline in hardwood flooring use, and the rapid increase in house size, which required larger amounts of softwood dimension lumber for framing."[64] On average, house size increased from 1,900 square feet in 1980 to 2,500 square feet in 2006.[65]

Our bigger houses require more timber. In his 2003 speech in Redmond, Oregon, following a series of severe fire seasons, President George W. Bush assured listeners that the Healthy Forests Initiative would, among other perceived benefits, "help our homebuilders." He noted that "good forest policy can be the difference between lives surrounded by natural beauty or natural disaster." He did not add that what he called "thinning the underbrush" also included logging, which could be incompatible with that "natural beauty," a term that reflects the romantic nationalism of both the nineteenth century and the 1960s.[66]

"Natural Beauty" in a Big Two-Hearted Nation

In January of 1965, pushing the Highway Beautification Act, which he would sign later that year, President Lyndon Johnson asserted in his State of the Union address that "for over three centuries the beauty of America has sustained our spirit and enlarged our vision."[67] The phrase "Natural Beauty"—the Johnson White House's catchall term for the environment—reveals how deeply nineteenth-century aesthetics informed public perceptions of the environment. What was considered worthy of protection was that which was conventionally scenic. The following May, Johnson addressed one thousand delegates at the White House Conference on Natural Beauty. As Frederick Jackson Turner had in 1893 credited the frontier with shaping what he believed was a uniquely American character, Johnson professed that the nation's beautiful landscape made the people who made the great nation. "[Natural Beauty] shapes our values. It molds our attitudes. It feeds our spirit, and it helps make us the kind of men and women that we finally become," Johnson contended, blending Turner's theories on the frontier and American character with Emersonian transcendentalism and eighteenth- and nineteenth-century notions of the moral landscape. Johnson added, "And the kind of men that we finally become in turn makes this great nation."[68]

The two-hearted vision of the landscape as both sublime panorama and cache of resources is a holdover from the Manifest Destiny era, but the year 1965 frames a few key events that reveal conflicting ideas about landscape. Imagine President Johnson in 1965, delivering this speech resonant with mythology grounded in nineteenth-century landscape aesthetics. Then picture the following: eleven of the twelve 1965 issues of *American Forests*, the magazine of a conservation organization founded in 1875, featured American landscape scenery on the front cover, while all twelve of the magazine's back covers carried an advertisement for the Homelite XP-1000 Direct Drive Chain Saw. That April, as the White House was gearing up for its Conference on Natural Beauty, *American Forests* featured an ad from the Reuel Little Tree Injection Company with the headline, "Kill Worthless Trees This Quick, Easy Way!" The ad featured a visual of a man plunging a lancelike implement into the roots of a tree. "Makes it easy and economical to kill trees in timber or kill worthless trees in pastures," the ad promised. "Sure kill every time."[69]

That September, one month before Johnson signed the Highway Beautification Act, the Forest Service announced the largest timber sale in its history when it opened bidding on 8.75 billion board feet of Alaska's Tongass National Forest. At the time, "a Forest Service spokesman in Alaska labeled those in opposition [to the sale] as 'extremist conservationists.'"[70] And throughout 1965, timber companies harvested over 6 billion board feet from our public lands in the western segment of the Pacific Northwest—the highest level since 1950, the first year for which records are available.[71]

The same month that Johnson championed Natural Beauty in his State of the Union address, a writer for *American Forests*, William B. Morse, admitted, "A West Coast clear-cut logging area looks like—well, say it, it looks like the devil"; the big slash "remains there as black, ugly litter until the new forest hides it." But new equipment from Crown Zellerbach was "a major breakthrough in timber harvest methods that may have far-reaching effects on forest utilization in the Pacific Northwest," he proclaimed, "and will affect all forest by-products, from scenery to wildlife."[72] Even as Johnson was hawking Natural Beauty, scenery and wildlife were, in Morse's words, "forest by-products"—commodities, and of secondary value at that. Morse was writing in the tradition of Moon and Brown, who had concluded fifty years earlier in *Elements of Forestry* that "the other ends served by the forest . . . are incidental." Crown Zellerbach's breakthrough was a machine called the Utilizer, "a small barker and chipper mounted on a large, low trailer with its own self-contained power plant," that would gobble up the slash and clean up the view after a clearcut—an industrial mode of landscape design.[73]

"Our artistry worked toward a model whose central image was the machine," William Kittredge recalls of his ranching childhood in southern Oregon. He adds, "We shaped our piece of the West according to the model provided by our mythology, and instead of a great good place such order had given us enormous power over nature, and a blank perfection of fields."[74] Technological power over nature remained the model in the mid-1960s, when Kittredge's childhood was well behind him. For all the cultural prattle about "Mother Nature," nature was still portrayed as humankind's enemy, and although such images would fall out of fashion in the following decades, ads for lawn products at that time portrayed homeowners at war with their yards.[75] Since World War II, we had been drenching our landscape in DDT, a pesticide that was labeled "the atomic bomb of the insect world."[76]

In 1965, as Morse was trumpeting the Utilizer, the Atomic Energy
Commission exploded thirty-nine nuclear bombs underground at the
Nevada Test Site and one in Alaska, but power over nature extended
only so far; sixteen of those underground blasts released radiation into
the air.[77] Considering the accelerating heat of both the Vietnam and
cold wars, it is hardly surprising that the industrial approach to forested ground now seems military in force. Such mechanistic forestry
operations like that afforded by the Utilizer offered a "blank perfection" of earth, stripping it of diversity to make way for the planting of
the next uniform crop of trees—if the cut would be replanted at all.

Political Ground

Mark Harmon, lead scientist at the Andrews Experimental Forest,
says that such extensive "clean up" strategies were common over the
last several decades, but were conducted based largely on assumptions
about "sanitation, productivity, fire protection, and logger safety,"
and fear of insect infestation, all of which "often had no basis in fact."
We are learning to think differently about what falls to—or what is
left or placed upon—the forest floor. Harmon, whom Fred Swanson
calls the Andrews's "head rotter," is directing a project focused on
the study of nutrient cycling of woody detritus—a field that Harmon
calls "morticulture." "The timber resources of the Pacific Northwest
were initially regarded as limitless," Harmon notes, and the region's
early twentieth-century loggers harvested only the largest, most valuable trees, often leaving stumps as tall as three or four meters. "In
1910, the typical harvest of an old-growth stand would have retained
65 percent of the live woody organic matter aboveground as slash,"
Harmon remarks—an amount he likens to the leavings of a "cata-
strophic fire or windfall."[78]

As timber became "scarcer," Harmon adds, "utilization standards
that removed only the 'best' volume became viewed as wasteful," and
by the 1940s remaining stumps were only 0.6 meters high and loggers
were taking thinner trees as well. "The earlier harvest practices were
viewed as wasteful, and therefore woody detritus became the symbol
of that waste," Harmon observes. If the stuff could not be sold, it was

destroyed or removed. "Woody detritus was not to be tolerated even if it cost a great deal of money to remove it," he notes, adding that the excessive clean-sweeping of cutover land "represented waste removal carried to its illogical extreme."[79]

In 1985, Harmon began a two-hundred-year study of the nutrient cycles of forests, or what the environmental writer Rick Bass calls the "tithing of rot," to determine "how much woody detritus is required to sustain ecosystem functions."[80] Morticulture, Harmon says, "emphasizes the culturing of something, in this case woody detritus. . . . [I]nstead of the type of logs to be harvested, it would deal with methods to produce woody detritus structures for ecosystem function."[81] At the Andrews, my colleagues and I visited one of Harmon's six log decomposition sites, this one located beside a gravel road above Lookout Creek, where Swanson picked up a chunk of Pacific fir and squeezed it; the wood gave in his hands, and then sprang back into shape with the elasticity of a sponge. Five-hundred-year-old Douglas firs and western hemlocks sheltered Pacific yew, a species of tree that Swanson calls "the druid of the forest with its cloak of mosses and modest stature after several centuries of growth."[82]

Amid this cacophony of green, harvested logs lay rotting where Harmon and a crew of loggers had arranged them when this study commenced in 1985. More than five hundred such logs of Pacific silver fir, western hemlock, Douglas fir, and western red cedar were harvested from the Andrews's cutting areas and placed strategically at the six decomposition study sites. The stories of rot roil beneath the fecund muck and within massive logs upholstered in moss.

"I like to imagine that each different tree, after it has fallen, gives off a different quality of rot—a diversity even in the manner in which nutrients are released to the soil," Rick Bass reflects of the forest around his Montana home. "The slow rot of a giant larch having a taste to the soil, perhaps, of bread; the faster disintegration of ice-snapped saplings tasting like sugar, or honey. The forest *feasting* on its own diversity, with grace and mystery lying thick everywhere."[83] The rot-monitors have determined that such poetic musings have a scientific foundation. Decomposition is complicated, Harmon says, involving factors such as "the chemical and physical nature of the wood, to decomposers involved, to the environment at the micro- and macro-levels."[84]

Harmon and his fellow researchers track rot with an eye toward understanding how time, conditions, and physical layout affect decomposition of woody detritus. If we track the forest's decomposition, then we can better understand what the ecosystem needs in order to sustain itself, and that could have important forest management policy implications.

Woody detritus is volatile political fodder. Oregon, like other western states, has endured devastating fires in recent years. In early 2006, a few Oregon State University graduate students caused a national flap when the journal *Science* published their research, which suggested that "forests recover from blazes more quickly if left alone to self-regenerate" than if they are logged, and that logging could even increase the risk of future fires.[85] Several Oregon State professors attempted to delay publication of the article, and the Bureau of Land Management, which had funded the students' research, revoked their funding amid industry complaints that the study was a politically motivated effort to stop White House–backed logging of Oregon's charred public lands. The students' grant was later restored.[86]

And anyone who has visited Alaska or the western lower forty-eight over the last decade or so surely has noticed the swaths of dead and dying trees, claimed by infestations of various beetles. I well remember encountering the relics of beetled trees and logs while hiking in Montana's Bitterroot National Forest. Inscribed with lacelike calligraphy or studded with holes that the beetles left in their wake, they struck me as beautiful. Some insist that beetle-killed trees should be logged to protect the surrounding forest. "Unless diseased or insect-infected trees are swiftly removed, infestations and disease can spread to healthy forests and private lands. Timely active management actions are needed to remove diseased or infested trees," states the Healthy Forests Initiative.[87] Beetle colonies are normal, and forest ecologists have uncovered signs of previous infestations.[88] But "the size and scale of these events are larger than we have seen in recent years," reports the environmental writer George Wuerthner.[89]

The current massive beetle outbreaks may even be caused, in part, by human behavior. Although it is not certain that climate change is causing the beetle infestations in western Oregon, scientists attribute rising beetle infestations elsewhere in the West to global warming.[90]

Warmer temperatures mean that the beetles are pushing northward and that more beetle eggs survive the winters.[91] Wuerther notes that "forest practices such as clear-cutting have regenerated large stands of vulnerable same-age, single species trees."[92]

Although his studies at the Andrews's log decomposition sites are focused on woody detritus rather than beetle kills, Mark Harmon says that "we need to develop a long-term, broad-scale view that is dynamic and that includes everything from [live trees] to snags to logs to highly decomposed material that functions as soil organic matter."[93] Harmon's log decomposition sites harbor the energies at work in the forest, transcribed in cipher by insect, plant, rock, weather, animal, fire, and earth. Harmon and his colleagues periodically cull samples from the rotting logs by slicing off sections, or "cookies," that they study like serial installments in a two-hundred-year story, decoding the arc of each log's unraveling. "Beneath the silence of moss / a mute voice / begging for voice," writes the Tlingit poet Robert Davis.[94] As I read over Harmon's words, I think of the stories murmuring beneath bark and moss; of Aldo Leopold sawing through layers of landscape history as he cuts down a tree in *A Sand County Almanac;* of the beetle glyphs whittled into western trees; and of the lost Pinchot paintings, an archive of America's intertwined ideas about art, nature, and nationalism, dumped in a hole in the Pennsylvania dirt.

The Bush White House insisted that the Healthy Forests Restoration Act would protect forests from wildfires, but coastal Oregonians are understandably conflicted about logging. Old-growth forests once covered much of their region. "By the 1950s much of northwestern Oregon's old growth had been consumed," writes David Peterson del Mar. "Logging and milling moved east and especially south, to the richly forested hills of Lane, Douglas, and Coos counties."[95] The timber industry suffered its own boom-and-bust cycles sparked in part by a "shortage of timber, international competition, fluctuations in demand, [and] mechanization," del Mar observes. He adds that "the thousands of loggers and mill workers thrown out of work in the late 1980s usually focused their ire on just one of these many interlocking factors: the northern spotted owl," which suffered precipitous population declines as more and more of its habitat was logged.[96]

Oregonians have learned the hard way that clearcutting endangers

more than owls and their habitat. In one single night in 1996, following heavy rains, five people were killed by two separate mudslides
attributed to extensive clearcutting.[97] The Andrews's Fred Swanson
had determined as early as 1975 that "clearcutting triples the chances
of landslides," which "are concentrated in the first 20 years following
logging, after old root structures have decayed but before new ones
can develop."[98]

Clearcuts are one of our nation's more striking symbols of environmental violence, and I passed several on my way to the Andrews. I also
passed farmlands lush with summer growth, and I thought of all the
miles I had traveled through the nation's agricultural flatlands during
several cross-country drives. Why did I find the clearcuts so shocking
when I never blinked at the miles of agricultural fields except to wonder which crops were grown there and for what purpose? Why did a
clearcut upset me more than the thirsty monoculture of a golf course
or a typical lawn?

Perhaps we react more strongly to clearcuts than to golf courses,
lawns, and big agricultural operations because, to eyes raised on
American landscape aesthetics, the scale of the devastation from a
clearcut is so much more dramatic. Some of us learn to vilify logging,
even though we may use timber-derived resources in our everyday lives.
Americans, in general, are culturally conditioned to appreciate the lawns
of both golf courses and homes; lawns are artifacts of the eighteenth-
century British aesthetic of the beautiful, and in the United States, we
associate tidy lawns with a pastoral democracy, community stability, and
even patriotism.[99] The agro-industry is a major polluter that has nothing to do with the pastoral republic that Thomas Jefferson envisioned,
yet agricultural fields are classic Americana. Small family farms, most of
which struggle to stay in business, are often represented as wholesome
and down-home American. Hector St. John de Crèvecoeur, in *Letters
from an American Farmer*, taught us that agriculture is holy work. "What
is a farm but a mute gospel?" Emerson mused in *Nature*.[100] Even the
word "pastoral" entangles farming with religious ministry. So agricultural fields don't shock the sensibilities in the same way that clearcuts
do. But to many Americans, mountains are holy ground.

How Forested Mountains Became Sacred Ground

We have an ancient tendency to think of mountains as sacred. The biblical psalmist lifted his eyes to the mountains for help and sought sanctuary there at the feet of God. Moses received the Ten Commandments at Mount Sinai, and Jesus delivered the Beatitudes and the Lord's Prayer during his "Sermon on the Mount." Yosemite National Park has its own Cathedral Peak, so named by the California Geological Survey in 1863.[101] And in 1875 the public enthusiastically welcomed Thomas Moran's *Mountain of the Holy Cross* (see fig. 1).[102]

American mountain ranges also gave our nineteenth-century predecessors some of their first national bragging rights. "In other parts of the globe are a few mountains which attain a greater altitude than any within our own limits, but the mere difference in height adds nothing to the impression made on the spectator," William Cullen Bryant exulted in the introduction to Appleton's 1872 *Picturesque America*. "Among our White Mountains, our Catskills, our Alleghanies [*sic*], our Rocky Mountains, and our Sierra Nevada, we have some of the wildest and most beautiful scenery in the world," he boasted.[103] Mountains like those Bryant described are aesthetically sublime—the province of God.

Most contemporary Americans may regard "tree huggers" as extremists, but the notion of trees as sacred beings has ancient roots. The Druids "revered" oak, yew, ash, and hazel.[104] Trees also figure prominently in the Judeo-Christian story of the universe: The Tree of Knowledge, at the center of the Garden of Eden, was the linchpin in humankind's fall from grace. In the Gospel of Luke, Zacchaeus sought enlightenment by climbing a sycamore to get a glimpse of Jesus, who was later crucified on a cross hewn from a tree.[105] "The groves were God's first temples," and "the gray old trunks . . . high in heaven / Mingled their mossy boughs," William Cullen Bryant wrote in his 1825 poem "A Forest Hymn."[106] Ralph Waldo Emerson called forests "plantations of God," where he found "sanctity," "perpetual youth," and "reason and faith." Emerson attained his meditative states "in the woods," a phrase that he repeats three times, like a prayerful refrain, in his famous "transparent eyeball" passage in *Nature*, suggesting that he could attain such states only among trees.[107] Trees were Henry David Thoreau's "shrines."[108] In *Walden*, Thoreau asserts, "I went to the woods because I wished to live deliberately, to front only the essential facts of life," a statement that implies he would find those "essential facts" only in a forest.[109]

In *Landscape and Memory*, Simon Schama notes that the Yosemite Valley's giant sequoias, which miners stumbled on in the 1850s, were taken to be "the botanical correlate of America's heroic nationalism at a time when the Republic was suffering its most divisive crisis since the Revolution." This is where American forests got tangled up with American civil religion and the empire-building project of Manifest Destiny that was freighted with Christian ideology. Schama adds that the nation eagerly adopted "the pious notion that the Big Trees were somehow contemporaries of Christ." The sheer size of the western trees inspired the public's awe. "The sequoias seemed to vindicate the American national intuition that colossal grandeur spoke to the soul," Schama observes.[110]

For William Cullen Bryant, the size of the trees seemed to speak directly to, or of, the national ego. The United States did not have the Old World's ancient ruins or its long history of western culture. "But Switzerland has no such groves on its mountain-sides, nor has even Libanus, with its ancient cedars, as those which raise the astonishment of the visitor to that Western region—trees of such prodigious height and enormous dimensions that, to attain their present bulk, we might imagine them to have sprouted from the seed at the time of the Trojan War," Bryant enthused in *Picturesque America*.[111] John Muir believed that no visitor to Yosemite "could escape the Godful influence of these sacred fern forests."[112] "The hills and groves were God's first temples," he wrote, echoing Bryant, "and the more they are cut down and hewn into cathedrals and churches, the farther off and dimmer seems the Lord himself."[113] Muir's reverence and language continue to inform environmental discourse; in the late 1990s, one environmental writer referred to logging in Alaska's Tongass National Forest as "cutting the cathedral."[114]

Many of us still cherish the notion that nature heals all hurts suffered in body and spirit. It's a story that comes down to us from ancient times, from peoples who still collect their medicines from the earth instead of the pharmacy, from the tradition of Rousseau's noble savage, from the poems, stories, and essays of British and American Romantics, and from countless stories and films. In Ernest Hemingway's "Big Two-Hearted River," a soldier returns from war, goes camping in a favorite forest haunt, suffers the POW's postwar flashbacks, makes love to a river while fishing, catches trout, climbs up the riverbank, lights a cigarette, and ambles away, cleansed by his immersion in the river's erotic waters. Everything, we know, will be more or less okay for the veteran after that riparian honeymoon.

"The living of life, any life, involves great and private pain, much of which we share with no one," observes the environmental writer Barry Lopez. In wilder places, he says, "you can hear your heart beat."[115] Perhaps there is some truth to the belief that nature can soothe the mind. Once, shortly before I moved out West, I found myself alone at my parents' home, and I wandered the yard because I could not stand to be inside the house, a structure in such ill repair that it seemed a metaphor for its occupants. The yard brought little relief; all of the fruit trees were strangling with fungus. To calm myself, I instinctively picked up a pine cone and held it to my nose. I expected pitch, but an entire forest drifted up and I stood there whole moments, inhaling the cool breath of earth.

The Sheltering Forest

By now it's common knowledge that the cool breath of earth is warming up. Clark S. Binkley of the Hancock Timber Resources Group, a timber investment management organization, notes that "trees produce one of the only technologies available to reduce atmospheric levels" of carbon dioxide.[116] Fred Swanson, Mark Harmon, and their colleagues at the Andrews Experimental Forest have concluded that the aggressive forest practices of the second half of the twentieth century "have released a substantial amount of carbon dioxide into the atmosphere." But they have also found that the Pacific Northwest could be "a significant, national-level carbon sink." "By altering forest harvest practices, increasing the length of time between harvests, and reclaiming former forestlands these studies indicate current carbon stores could be almost doubled without reducing the volume removed for forest products," they note of what sounds like a very promising scenario.[117]

Clearcutting is incompatible with carbon storage and at the heart of so much of the controversy that surrounds logging. My first visit to a clearcut came when Fred Swanson piloted our van into one on the edge of the Andrews Experimental Forest, and my first thought was that he had driven us into a graveyard. The gray stumps looked like headstones in the frowsy growth. This impression, I would learn, is fairly common; several members of our party had the same reaction, and Thoreau thought the same thing when he ambled into a meadow in 1858. "This new pasture, with gray stumps standing thickly in the now sere sward, reminds me of a graveyard," he wrote in his journal. "And on these monuments you can read each tree's name, when it was born (if you know when it died), how it throve, and how long it lived, whether it was cut down in full vigor or after the infirmities of age had attacked it."[118] This forest, Swanson said, had been cut twice—first fifty years ago, and again five years ago. Wood thrush notes, part washboard, part xylophone, rang through the forest at the perimeter of the cut; robins warbled closer by. Fragments of wood and bark were scattered like bones among daisies, irises, nettles, tangles of multiflora roses, and green plastic shotgun shells. The place was hardly dead, but neither was it a forest. But it was lush compared to the clearcuts I would see later, when I headed toward the coast.

"Common Ground"

The most encouraging thing about the Andrews Experimental Forest and the other Long Term Ecological Research Sites is that they exist at all, and that, in a culture that places ever greater value on short-term rewards, in these places we are stepping back and thinking long-term. As a geologist, Fred Swanson is accustomed to taking the long view and to monitoring change in its dramatic manifestations, such as the explosions of Mount St. Helens, and in the nearly imperceptible gradations that water and wind work into rock. The view at the Andrews is so long that the scientists working there now will not live to complete all of their experiments, some of which will extend over two hundred years. The Blue River Face Timber Sale Unit is part of one of ten Adaptive Management Areas of the Northwest Forest Plan, where scientists study methods of conserving wildlife habitat while meeting a variety of social needs. Adaptive management is a way of learning by doing, Fred Swanson says, by implementing "management practices, monitoring their effects, and adjusting plans . . . based on lessons from monitoring, operational experience, new research, and other sources," and adjusting policy according to what we learn as we go.[119]

The forest in this unit appears to have survived at least two fire regimes, Swanson says, so that what remains is "a mixture of low density old growth" and "mature forest," meaning that 400- to 500-year-old Douglas firs are interspersed with 140-year old Douglas firs, western white pines, red cedars, and hemlocks.[120] "As ecosystem science has developed over time, forests have come to be viewed more as ecosystems than as tree farms," Swanson says.[121] The work carried out at this unit is based on what Swanson alternately calls "landscape dynamics" and "ecosystem dynamics," based on an understanding of historic levels of disturbance—such as fire or logging—in a particular area.[122] By the time I saw the Blue River Face Timber Sale Unit, it was five years into the plan and had been thinned fairly dramatically. Teams of workers had removed logs along a skyline cable to minimize damage to the ground, leaving a substantial buffer zone along the stream.

Foresters had also partially burned the site in order to create snags for wildlife habitat. Prescriptions for this site call for logging at intervals of 180 and 260 years, leaving, respectively, 30 percent and 15 percent of live trees in place. "Not all logged areas had that confusion of spirit or loss of grace; some of them retained, or reshaped, the grace of the woods (or rather, the grace of the woods altered itself and still flowed around and through those areas that had been logged with care and respect)," Rick Bass writes of the logged areas around his Montana home.[123] Bass could have been writing about this parcel of forest. From my decidedly unscientific perspective, it looked like good practice, with room for trees to grow and enough cover for animals to move about. But Swanson observes that with this level of cutting, not everyone would consider it a forest.

Swanson points out that Americans have traditionally approached forestry with conflicting worldviews. An anthropocentric worldview would support "intensive plantation farming," whereas a biocentric worldview "may consider species protection paramount and support management based on a conservation biology approach." A third approach seeking to balance the needs of various groups by "emphasizing ecosystem dynamics may favor sustaining the range of historic conditions as a management approach for balancing commodity and species objectives."[124] In order to meet the needs of the local community, he adds, "we probably need such a blending of these and other worldviews, including indigenous knowledge, to arrive at sustainable policy. The multiplicity of societal objectives—sustain species, ecosystems, ecological services—calls for a multiplicity of approaches."[125]

"It is humanity's collective sense of values that has guided our interactions with forests through history," write Constance Best and Laurie Wayburn, founders of the Pacific Forest Trust. "And those values change. What was once seen as 'good' (carving a field out of the wilderness for a family's home) may now be seen as 'bad' (cutting the last old growth, or fragmenting a forest habitat)."[126] Best and Wayburn could be writing of Sanford Robinson Gifford's *Hunter Mountain, Twilight*, the painting that hung above Gifford Pinchot's fireplace. Values depend on one's perspective, and our collective values regarding public lands are conflicted. But "common sense" and "community" are among those things Americans value most, according to the pollster Frank Luntz.[127] Both terms hearken back to our nation's creation story. John Winthrop, governor of the Massachusetts Bay Colony, preached visions of a Christian "community" before the *Arbella* came ashore, and Thomas Paine exhorted American colonists to exercise "common sense" by freeing themselves of the yoke of British rule. "Common sense" and "community" figure repeatedly in Bush White House documents promoting the Healthy Forests Restoration Act. The Healthy Forests Initiative invokes "community" in its subtitle: "An Initiative for Wildfire Prevention and Stronger Communities."[128] But the meaning of "common sense" and "community" also depend on one's perspective, and in the case of the Healthy Forests Restoration Act, "community" does not broadly involve the people to whom public lands belong. The Healthy Forests Initiative promised to "improve regulatory processes" and "expedite implementation of fuels reduction and forest restoration projects" by avoiding the current "needless red tape and lawsuits [that] delay effective implementation of forest health projects."[129] That sounds good, but what the plan effectively does is stifle the community by curtailing opportunities for informed public comment on "healthy forest" logging projects.[130]

I don't pretend to know what's best for our forests in Oregon or elsewhere. But I think it's vital to sift through the layers of carefully wrought "spin" to uncover the realities buried beneath the rhetoric and the pretty pictures.

"Plantation forestry has been half a century in development, conservation biology is a few decades old, and landscape management based on concepts of ecosystem dynamics is in its infancy," Swanson notes.[131] It may take a while for the public to embrace such an approach. "We have to talk about compromise and common ground," Swanson says. "When you get to the commons, that's where you've got to park your firearms at the door."[132]

Clearcuts and Scenic Corridors

When I left the Andrews, Kathleen Dean Moore advised me to head west on Highway 126, then south when I got to the coast. On that route, she said, I would find the landscape that I really needed to see.

"Landscape is more than a passive reflection of a nation's civil religion and symbolic totems," writes the geographer Kenneth Foote. "Landscape is the expressive medium, a forum for debate, within which these social values can be discussed actively and realized symbolically."[133] As I drove along western Oregon's highways, it became painfully clear that we need to reconsider both "social values" and the ways that we choose to represent our environment. Toward the coast, cattail and wildflower wetlands gave way to dense conifer stands that gave way to clearcut mountains, strewn with slash or stripped down to dirt. I passed scalped mountain after scalped mountain. From the road, so many of these torn-up mountains appeared devoid of vegetation. Others were replanted with grids of young trees that, except for the symmetry, reminded me of the scrub vegetation that covers the desert mountains around Reno. This, I knew, was an illusion that

would disappear; the mountains would lose their pincushiony look as
the ground cover and understory crept back among the grid of trees.

From my car, I saw the aesthetics of national narratives played out
on a local scale on a landscape marked by American iconography.
When I reached the town of Veneta, I found several national icons of
American progress converging at the intersection of Highway 126
and Territorial Highway. We were just a few days from the Fourth of
July, and a commercial for TNT Fireworks blared from the radio—a
reminder that I could show my patriotism by celebrating Independence
Day with colorful explosives. To my left stood a little red barn with
white trim, shiny and new but classic in form and color—a polished
pastoral icon. In a strip mall across the intersection, another barn—
this one a Dairy Queen—kept company with a Chevron station, with
its red, white, and blue insignia. An American flag the size of a bill-
board waved above the Siuslaw Bank, named after one of the regional
peoples forced out of place when white settlers arrived. Straight ahead
of me, due west on 126, stood a denuded mountain, its pelt of forest
logged out in a massive clearcut. There were many more.

The numbers and scales of the clearcuts along Oregon's coast are
simply staggering. Imagine entire cities with everything blasted away
except for the foundations of the buildings, and you can get the idea
of the scale of the impact. Many of the mountains looked completely
gouged, and that, I figured, was silviculture: the trees were harvested
and the earth scraped raw.

Sometimes the road-level clearcuts were partially veiled behind a
screen of leaves, which in some places provided about as much cover
as a strip of green gauze. Like makeup painted on the face of a corpse,
the timber fronts are designed to mask the ruined forest behind a pre-
sentable facade. They reminded me of the "veneer of dirt" the former
Oregon logger Robert Leo Heilman says is left on the ground after
clearcutting.[134]

There is an irreconcilable tension built into our conflicting myths
about our landscape. If we envision the United States as a kind of
Eden or Canaan, as many of our predecessors did, and if we embrace
the teleological promises of Manifest Destiny, then we must regard
our natural resources as provisions doled out to a fortunate nation.
In this view, development is a kind of divinely ordained utilitarian-

ism—even a sacred obligation. Yet if we also embrace the aesthetic of the sublime, then we think of mountains, with their big trees, as God's domain. And if we imagine our nation in terms of its "purple mountain majesties," then clearcutting stabs at the heart of American patriotism—our civil religion. In the 1970s, speaking in the value-laden vocabulary of popular discourse, Senator Gale McGee of Wyoming "called clearcutting a 'shocking desecration that has to be seen to be believed.'"[135]

Desecration or not, the senator was right; clearcutting *does* have to be seen to be believed. After an hour or so of driving my eyes grew accustomed to Oregon's weird geometry, and I saw that nearly everywhere I looked, portions of the mountains had been clearcut at some point and were in various stages of regrowth. The bigger shock came later, when I pulled off the coastal highway onto Douglas County Route 49, a public road that also serves as a logging road. Everyone should see such places. Whole forests had been amputated from the slopes of the mountains. Gifford Pinchot denounced wartime clearcuts as "tree butchery," and these mountains looked critically wounded.[136] As far as I could see, whatever was not ripped apart had been slashed at one point or another. I drove ten miles until I reached the point where the road was closed to the public, and it was the same horrorscape all the way in. I drove back out, and near the top of the hill I got out of my car and looked around at the ruined slopes. Timber depredation.

I have lived too long in cities, where it's easy to forget about the forests, loggers, and timber mills that feed my paper-hungry lifestyle. Pacific Forest Trust's Best and Wayburn cite the "need to gain greater public understanding of the contributions of private forests to our lives and society. The growing lack of connection between people and forests in our urbanized society is itself a key barrier to increased public and private investment in forest conservation," they write.[137]

Like any industry, corporate forestry has its own esoteric lexicon. The language reduces—on paper, at least—acts of environmental violence to abstract processes. "Although tree planting is part of something called reforestation, clear-cutting is never called deforestation—at least not by its practitioners," Robert Leo Heilman muses. "The semantics of forestry don't allow that. The mountain slope is a 'unit,' the forest a 'timber stand,' logging is 'harvest' and repeated

logging 'rotation.'" And clearcuts, Heilman adds, are designated by
the phrase "Overstory: Zero."[138]

"You do violent work in a world where the evidence of violence
is all around you," Heilman writes of logging.[139] I stood for a while
looking at the Overstory: Zero. Aerial maps of this region show what
looks like an earthen jigsaw puzzle bereft of several pieces, mountain
after mountain gnawed down to bone. On the ground, it looks like
patchwork madness.

"A place is healthy if it has cores of wildness in it," observes Rick
Bass.[140] Clearcuts are more than eyesores, and that patchwork logging
poses a threat to biodiversity because it removes those "cores of wildness." David Lindenmayer and Jerry Franklin, two highly respected
forest scientists whose combined careers span more than seventy years
in the profession, write that "clearcutting provides hostile conditions
for the movement of many organisms." It is difficult for wildlife to
thrive even in forest reserves if the terrain around them is clearcut;
animals crossing a clearcut between forested segments have little or
no cover.[141] Henry David Thoreau was more than one hundred years
ahead of the science of biodiversity when he observed that "in wildness
is the preservation of the world."[142] Forest management, Lindenmayer
and Franklin say, "will require a far more comprehensive and multiscaled approach than simply partitioning forest lands into reserves
and production areas."[143] They believe that "slowing the rate of clear-
ing is critical for biodiversity conservation," and they advocate the
establishment of forest reserves and management plans for adjacent
forestland that will help conserve biodiversity.[144]

One of the most shocking things about the road I stood on is that
portions of it run roughly parallel to Route 38, a scenic corridor complete with a wildlife viewing station. The forests were shredded for
miles around, yet if I had climbed to the top of the ridge and looked
down the other side, I would have seen forested slopes, the Smith
River, a corridor of green bisected by a highway, and Roosevelt elk
grazing or resting on the grass. The tourists driving along this road,
pausing to take pictures of the waterfowl and elk, have no clue about
the clearcutting on the opposite flank of the ridge.

"On the Oregon coast, the children know mostly fish-poor, flood-
stripped streams. Here, all estuaries are fouled, and no river water is

safe to drink. That's the way it is. Why should they think it could be any different?" writes Kathleen Dean Moore. She adds, "It's not just their landscape that has been clear-cut, but their imaginations, the wide expanse of their hope."[145]

It's pure fantasy to imagine that my life would have been much wilder in western Oregon than in eastern New Jersey. The same year that I was kicking balls on New Jersey streets and the White House was pushing Natural Beauty, we were slicing up the forests of the Pacific Northwest, and the legacy of my lifetime has left things far worse. Sixteen million salmon once shot from the Pacific ocean up the rivers "of California, Washington, Oregon and Idaho to spawn."[146] Now the salmon are gone from 40 percent of their traditional spawning grounds in the lower forty-eight states, a loss that the National Park Service attributes to "logging, grazing, agriculture, urbanization, channelization, and road-building."[147]

Ground Truth

When I first moved to the West, I lived in Missoula, Montana. Evenings, the wind rushing down Hellgate Canyon was fragrant with the sweetness of trees and earth—laden, I thought, with the mingled forest scents of the Rattlesnake Wilderness at the northern edge of town. Eventually, I figured out that the smell I so loved was the windborne fragrance of the mountains of woodchips piled up on the grounds of the Stimson Lumber Company, five miles upwind in Bonner. Some days the air in town was stagnant with the sewerlike stench of wet cardboard, carried east on the wind from Smurfitt-Stone's container plant in nearby Frenchtown. Those two aromas—mountains of sweet-scented woodchips, shitlike pall, both seasoned with smoke from summer forest fires—form the olfactory frame of my memories of Missoula. Logging trucks loaded with timber drove through town, each ringed log a scroll inscribed with the intertwined stories of its forest home and American timber consumption.

"Most of us would rather accept the idea of an enemy among us than examine ourselves for signs of the enemy within each of us," Robert Leo Heilman writes.[148] It was easy to diss Smurfitt-Stone and rag about the awful stench, easy to trash timber company Plum Creek for selectively logging a mountainside near the Millpond Dam, leaving the remaining trees looking like conifer pins in a pegboard. It was harder to look critically at the artifacts of my own timber-hungry lifestyle. But to pretend that I do not participate in the logging of American forests would be to manufacture my own pulp fiction.

The good news about northwestern forests—if only to conservationists—is that we are cutting fewer trees from our national forests. Since the spotted owl was listed as an endangered species and the Northwest Forest Plan was implemented in 1994, cuts on public lands in the Pacific Northwest have declined steadily and dramatically, down from 6.6 billion board feet in 1989 to 1.4 billion board feet in 2002.[149] The Forest Service no longer practices clearcutting in the range covered by the Northwest Forest Plan. But here's the rub: while the harvests from northwestern national forests have fallen, overall consumption of timber has not. The burden of supply has simply shifted from public to private lands, to other regions, and to imports. In 2002, the most recent year for which regional estimates are available, the western segment of the national forests of the Pacific Northwest supplied only 1 percent of our softwood lumber, but private forests in the same region provided 62.8 percent of the stock.[150] According to James L. Howard, in 2005, lumber imports to the United States reached a record high of 25.7 billion board feet. Eighty-five percent of these imports came from Canada.[151] The southern region of the United States provided the bulk of our lumber—"35% of all softwood lumber and 80% of all hardwoods" in 2005.[152] The Pacific Trust's Best and Wayburn report that "70% of all forestland in the South, and 67% of all forestland in the Northeast and North Central Regions," is in private hands. Private forest owners control 58 percent of all timberland in the United States.[153] Another rub: private forestlands are subject to state rather than federal laws, and regulations vary from state to state.

Nationwide, we lose one million acres of forestland annually.[154] "North Carolina, California, Florida, Georgia, Massachusetts, and Washington lead the country in forest loss," Best and Wayburn report.[155] That means that owners are harvesting the trees and then selling off or otherwise developing the land, so what was once a forest becomes, say, a housing development. According to the Forest Service's Pacific Northwest Research Station, "after a forest is converted to urban uses, the ecosystem services, such as water and air filtration, biodiversity protection, and carbon storage, are effectively gone."[156] Ralph Alig, a Forest Service land economist, reports that "more than 50 million acres of U.S. forests are projected to be converted to developed uses (e.g., parking lots) over the next 50 years, as the U.S. population grows by more than 120 million people."[157]

Forest loss is a worldwide problem. "We are greatly concerned that approximately half the world's native forests have been cleared in the past forty years," state Lindenmayer and Franklin. "It has been estimated that between 90 and 95 percent of the world's forests have no formal protection."[158]

Timber-Framing the Nation

Photos of the clearcuts like those along Oregon's coast never make it onto postcards or scenic calendars. More often, such artifacts feature landscape photos composed according to Manifest Destiny aesthetics: conquer the spectacular if we can or partition it if we must, use everything else, and romanticize representations of the landscape. Nor are clearcuts destinations on our scenic routes. But these, too, are our purple mountain majesties, butchered down to the dirt. If everyone could see clearcuts, would our nation still consume so much timber?

Angela Miller notes that by the mid-nineteenth century, "landscape art . . . was a cultural endeavor directed at consolidating a middle-class social identity bound up with the civilizing mission."[159] "The moral picturesque," she observes, "helped organize the meanings of landscape imagery by imbuing it with literary and religious content."[160] The "moral picturesque" was a visual construction of the "moral geography" that was invested in the landscape, evolving from the Puritan ideal of "plant[ed]" churches and towns to Manifest Destiny–visions of empire.[161] The opulent frames that encased landscape paintings often represented civilization's control over wild nature.[162] Today, writes Malcolm Andrews in *Landscape and Western Art*, perception or experience of landscape "is increasingly mediated by frames of one kind or another: the window, the camera viewfinder, the television set, the cinema screen."[163] Perhaps because he lives in the United Kingdom, Andrews left out the quintessential American frame—the windshield of the car or SUV through which so many of us view the world.

Tourists driving along Oregon's Pacific Coast Highway might remember that the surface of the ocean conceals multiple human-induced problems. And then we might look east toward the land, and recall that despite what our national myths tell us, forests are more than symbols of economic resources or "natural beauty." These wrecked forests are the logical outcome of the values and practices that Thomas Cole, painter of *The Course of Empire*, warned against in 1836.

Like the Las Vegas Strip with its perpetual neon carnival, the Pacific Coast Highway is classified as an All American Road, a designation the secretary of transportation determines based on one or more of the road's "archaeological, cultural, historic, natural, [and] recreational qualities."[164] With its startling juxtaposition of mangled nature and designated scenery, perhaps Oregon's stretch of the Pacific Coast Highway truly is "all American." It exemplifies conflicting ideas about land use and landscape representation—what gets used and what gets preserved, which features are represented as scenic resources and which are hidden away, the timber essential to the current American lifestyle, the scenery essential to American national mythology.

As I stopped at the various scenic turnouts and faced the ocean with my back to the clearcuts, it struck me that I was little different from the eighteenth- and nineteenth-century picturesque tourists who held up their Claude glasses and turned their backs on the land, the better to see an idealized version of it—tinted, balanced, and contained within the frame of the mirror. If I faced the ocean with the forest razed behind me, was I not choosing to see the world, as St. Paul said, but through a glass, darkly?[165]

4
Open (for Business) Range
Wyoming's Pay Dirt and the Virtual Sublime

In the open space of democracy, the health of the environment
is seen as the wealth of our communities.

Terry Tempest Williams, *The Open Space of Democracy*

On my first drive west to my new home in Montana, I was so happy
that I probably could have fueled my car with my own adrenaline. I
had left New York City behind, trading pastel suits and a skyscraper
job for jeans and a small town-teaching position; in a few more weeks,
I would toss out the lipstick, too. I pounded the horn for joy when I
reached Wyoming (*almost there!*), startling several horses in a roadside
pasture. A year later, on a circuitous trip along western back roads, I
cheered like a Yankee fan, howling and bouncing on the car seat when
I got caught behind a herd of cattle on the outskirts of Cheyenne.

On my most recent trip to Wyoming I drove east from Nevada on
a late spring day toward a landscape now as promising for its reserves
of coal, oil, gas, and uranium as it once was simply for its open space.
The earth bears the marks of our cultural currents, and on the ground
in Wyoming, we can trace the shifting patterns that the confluence
of aesthetics, industry, and public policy—powered by the rhetoric of
patriotism—continue to carve into the land as if they were branches
of a braided river. I wanted to see how my appetite for energy was
reshaping Wyoming's topography, and I was heading toward Pinedale,
a formerly quiet community in the Upper Green River Valley that has
become the natural gas industry's latest boomtown.

Gas Range

Unfold a map of Wyoming and you will find the language of the nation's expansionist era printed all over the state, with towns, counties, mountains, forests, and bodies of water bearing the names of artists, explorers, and emigrants, as well as the resources that settlers found there. Unfold a Wyoming newspaper and you will find the same expansionist-era language attached to many contemporary energy extraction projects. In July 2006, Dale Groutage, campaigning for a U.S. Senate seat, called Wyoming "the energy capital of North America."[1] That's a big burden even for a state that calls itself Big Wyoming. The nation's top coal-producing state, in 2008 Wyoming gave us more than 467 million tons of coal, 52.9 million barrels of oil, and 2.3 trillion cubic feet of gas.[2] Wyoming's legendary Green River Valley harbors some of the nation's most productive gas fields.[3] Framed by frontier rhetoric, the energy projects reflect attitudes toward landscape and land use that appear to have changed little since the decades immediately following the Civil War.

Angela Miller writes that works of landscape art served a nation-building role during the post–Civil War era as they helped to foster a sense of American identity and mission.[4] Nineteenth-century images of Wyoming figured prominently in articulating the relationship between Americans and their nation, and the Green River Valley and the Wind River Range to the north were already famous when painter Thomas Moran arrived in Green River in July of 1871. Trappers, mountain men, and Native American traders had converged at Green River for their rendezvous several times between 1833 and 1840, and explorer John Charles Frémont had waved an American flag atop a Wind River peak on his first of five expeditions through the West.[5] Following Moran's arrival in the territory, Wyoming would become the site of a watershed change in public policy. Moran's watercolor sketches and oil paintings of Yellowstone's sublime topography, along with William Henry Jackson's photographs and Ferdinand Hayden's geological report, so impressed lawmakers that in early 1872 they designated Yellowstone as our first National Park, marking the first time that a portion of terrain deemed aesthetically sublime was partitioned off as a park for the entire nation. Later that year, Congress purchased

FIGURE 4: Thomas Moran, *The Grand Canyon of the Yellowstone*, 1872. Oil on canvas, 84 x 144 in. Interior Museum, U.S. Department of the Interior, INTR 3001.

Moran's acclaimed oil painting, *Grand Canyon of the Yellowstone* (fig. 4) for $10,000 and hung it in the Capitol building.[6]

Moran's images of Yellowstone may have fired the public imagination, but his Green River paintings played to another kind of longing. Captivated by the pastel bluffs that surrounded the town, Moran painted multiple renditions of them over the next few decades. The bluffs are the most realistic elements in his Green River paintings, as the artist engaged in what his biographer Nancy K. Anderson calls the "wholesale erasure" of the town's industrial landscape. In place of an enormous railroad depot, Moran depicted bands of Native Americans on horseback, even though the artist saw no Native peoples in the area. Moran's Green River images "fed the lingering hunger for spectacular New World landscapes," Anderson writes.[7] Yet during this era, observes Merle Curti in his seminal examination of American patriotism, *The Growth of American Thought*, the "needs and values of business enterprise were intimately associated with patriotic and nationalistic ideas and sentiments."[8] Enthusiastic public reception of works of landscape art indicates how deeply ideas about landscape and

Manifest Destiny aesthetics were embedded in American visions of national identity, land use, and commercial prosperity barely twenty-five years after John O'Sullivan coined the term "Manifest Destiny." It also reflects conflicted longings for both unblemished terrain and a resource-rich Promised Land awaiting the cultivation of empire.

"By the mid-nineteenth century," writes Jonathan M. Hansen in *The Lost Promise of Patriotism*, ". . . liberalism so dominated the political and cultural landscape that most Americans had come to associate 'freedom with acquisition, liberty with property, and politics with strict non-intervention.'"[9] By the end of the nineteenth century, this consumer-oriented current of patriotism would be joined by a "version [in which] true patriots were often represented as male warriors," notes the historian John Bodnar. This bellicose ideal became even more powerful after World War II, Bodnar observes, adding that by the 1980s, this line of patriotism "offered spectacles of power and the veneration of elite groups of warriors."[10]

Contemporary energy projects frequently evince a strain of patriotism that appears to be a hybrid of these nineteenth- and twentieth-century currents. Cast in images and rhetoric that reflect Manifest Destiny aesthetics—conquer the spectacular if we can, partition off the spectacular if enough people insist, use everything else, and romanticize the landscape in pictures—they meld notions about terrain, commerce, technology, and war in a progressive narrative even as they damage the landscape that still figures so prominently in national mythology. I wanted to read the stories that my times are engraving into the ground in Wyoming, and I headed to Green River, fretting as I drew close that I might miss the freeway's exit sign for the town. I didn't miss the sign, but I didn't need it; I knew I had arrived at Green River because I recognized the sandstone bluffs from Moran's paintings.

I stopped first at Green River's Thomas Moran Overlook Park, where I lounged at a picnic table on a clipped lawn and regarded the bluffs that had so charmed an artist and a nation. In *Picturesque America*, the travel writer E. L. Burlingame, disheartened by Wyoming's sagebrush flats, saw Green River as "a welcome relief to the monotony that [had] marked the outlook during the miles of level desert that are past." Burlingame marveled at the landscape. "Here the picturesque forms of the buttes reappear . . . ," he wrote in 1872. "Fantastic forms abound everywhere, the architecture of Nature exhibited in sport," he marveled.[11] For Burlingame and his readers, the environment was an art museum or gallery, with objects "exhibited" to please the human eye; today the City of Green River calls itself "Nature's Art Shop."[12] Across the street from the park, a Walmart truck sat in front of the Oak Tree Inn next to a Sinclair station with its signature dinosaur, a reminder that my gas came from fossils and not simply from the pump. I noted the roar and bleat of traffic on the Interstate, the mingled smell of sweetgrass and sagebrush, and the park's electronic sign with its rolling digital announcements of upcoming concerts. On the far side of the river, a rail line followed the riverbank and the city sprawled across the flats and up into the hills. I got back in my car and pointed east, then north toward Pinedale.

Pay Dirt in the Promised Land

If Moran, Burlingame, Frémont, and the trappers could travel through the Green River Valley today, they would find themselves in what scientists call the Greater Yellowstone Ecosystem, a spread of land that includes Yellowstone and Grand Teton National Parks and the surrounding communities in Montana and Wyoming. They would also find a landscape transfigured by energy development. One hundred thirteen miles upstream from Moran's last stop on the train and barely two hundred miles south of Yellowstone, Pinedale—population eighteen hundred or so—is the biggest town in Sublette County. And Sublette is Wyoming's fastest-growing county; its population has swelled by 40 percent since 2000.[13] Pinedale sits along the southwestern flank of the Wind River Range and on the shoulder of public lands that include the two-hundred-thousand-acre Anticline and the adjacent thirty-thousand-acre Jonah Field, both administered by the Bureau of Land Management (BLM).

If Yellowstone and Grand Teton are among the National Park System's crown jewels, the Pinedale Anticline and Jonah Field are pay dirt to the energy industry. The Pinedale Anticline is checkered with more than 850 gas and oil wells, and several companies operate more than 600 gas wells in Jonah Field; thousands more are planned.[14] Jonah Field "produces 1.5 percent of the daily natural gas needs in the United States."[15] In aerial photos, Jonah Field looks like a city. Gas drilling in the Pinedale Anticline and Jonah Field also yields a profitable by-product—oil. *Casper Star-Tribune* reporter Jeff Gearino writes that Jonah oil is "a high-quality condensate" that producers can "[refine] into gasoline, diesel, and other (petroleum) products." Aggressive gas drilling and culling of the condensate oil have rendered Jonah Field and the Pinedale Anticline "Wyoming's top two oil-producing fields."[16]

Geologists working for EnCana, one of the area's main operators, describe the Lance Formation, which is the "principal reservoir" at Jonah Field, in language resonant with geological poetry. "The Lance Formation is comprised of braided to meandering fluvial sandstones intercalated with overbank siltstones and mudstones," they write, and I wonder if they sense the painterly lushness of their language even as they estimate the lavish stores of gas beneath the sagebrush.[17]

I turned off the Interstate at Rock Springs, home of the mammoth Jim Bridger Power Plant, named for the legendary mountain man and explorer whose name is all over Wyoming. Operational since the early 1970s, the Jim Bridger Power Plant is "one of the largest coal-fired power plants built in the West."[18] It is also Wyoming's biggest emitter of carbon dioxide, which is among the primary causes of global warming.[19]

I aimed north toward Pinedale and paused at the town of Eden, named by Mennonite settlers who had deemed it a "land of promise."[20] There are Edens, Paradises, and Canaans all over the United States, christened by Bible-packing emigrants who were hoping for a shot at re-scripting the story of the fall of humankind and who conflated notions of the first garden with dreams of the Land of Milk and Honey.[21] With its gas station and convenience store, this little Eden was the Land of Coffee and Snickers Bars, and it felt like a corner of Paradise. Paradise with trucks. Tanker trucks rumbled past, traveling south from Pinedale through country that appeared to be mostly ranchland punctuated by the occasional roadside ranch house.

After the White City: Romanticizing the Range

"Nobody ever lived like that," William Kittredge recalls his rancher grandfather telling him when he came across a paperback copy of *Ranch Romances*.[22] Even if nobody did, the romanticized ranch and rancher are American archetypes. The iconic rancher is the rugged cousin of Hector St. John de Crèvecoeur's farmer—a merging of Thomas Jefferson's pastoral ideal and Teddy Roosevelt's Rough Rider, charged with the task of shaping the frontier into Promised Land. The typical depiction of the rancher as male reflects a deeply entrenched cultural patriarchy. Ranches have long been family-based operations; the 1862 Homestead Act permitted unmarried women to file their own homestead claims; and in 1869 Wyoming was the first territory to grant women over twenty-one the right to vote.[23] The popularity of Thomas Moran's Green River landscapes indicates that the public was romanticizing life on that open landscape even before Frederick Jackson Turner delivered his 1893 eulogy for the frontier at the World's Columbian Exposition in Chicago. Constructed for the exposition, the electrified White City had "more lighting than any city in the country," writes the cultural theorist David Nye.[24] It was a dazzling manifestation of what Leo Marx calls the technological sublime.

Representations of the technological sublime, Marx writes, were framed by progressive rhetoric that supported the innovations of the expansionist era's industrial leviathans.[25] Richard Slotkin observes that visitors to the White City passed from the exposition's renditions of "the Wild West to the metropolis of the future . . . from displays of primitive savagery and exotic squalor to a utopia of dynamos and pillared facades."[26] Nye asserts that the White City "provided a visible correlative for the ideology of progress and abundance through technology."[27] The mighty sublime would power the grandest factories and illuminate even modest homes. With its blocks of monumental bright white neoclassical buildings and its stunning nighttime displays of electric light, the White City epitomized human mastery of two frontiers: the continental frontier subjugated by empire and the new frontier of American energy, made visible in what Nye calls the electrical sublime.[28] The White City offered spectacular forecasts of electrified urban landscapes of the future, even as the culture continued to produce images of a past of wide open spaces.

Although ranches operate throughout the United States, films and other media usually depict them in the sublime landscapes of the West, terrain that summons notions of the frontier and its presumed influence on the virtues that Frederick Jackson Turner ascribed to its inhabitants: "coarseness and strength," and the seemingly contradictory independence and community spirit that Turner insisted repeatedly strengthened American democracy. The rancher, though landed through lease or purchase, is twinned with the itinerant cowboy, romanticized in the visual arts of Charles Russell and Frederic Remington, in the novels of Owen Wister and Louis L'Amour, among others, and in twentieth- and twenty-first-century films and television programs. And the cowboy is inseparable from the open spaces of Wyoming; the Bucking Horse and Rider is a state- and federal-registered trademark of the Cowboy State.[29] When federal lawmakers designated July 22, 2006, as the first National Day of the American Cowboy, Idaho senator Mike Crapo observed, "The American Cowboy continues to symbolize the American ideals of freedom and fair play," a remark that indicates a commingling of ideas about lifestyle, landscape, and democracy with the mythical freedom of the frontier.[30]

One Big Neighborhood

Wyoming may be the Cowboy State, but it has long been both mining and ranch country. The state seal, established in 1893 and modified in 1921, features the words "Mines," "Oil," "Livestock," and "Grain," and the figures of a woman, who symbolizes "Equal Rights," and a miner and rancher.[31] Today there is some tension between the latter two of those old neighbors. Many of the ranchers in the Pinedale vicinity, alarmed by the rapid-fire development around their homeplace, believe that the extraction industry is encroaching on their freedom and that the play is not entirely fair. Since 1999, the number of roads has increased by 36 percent on the Pinedale Anticline and by

100 percent on Jonah Field, straining local wildlife communities.[32] Some residents feel that their traditional lifeways are threatened by burgeoning development, which pushes them out of their hunting grounds and grazing and recreation areas and compromises the integrity of regional ecosystems.[33]

Energy extraction on both the Pinedale Anticline and Jonah Field falls directly in the migration routes of big game, such as mule deer and pronghorns. Between 2001 and 2009, mule deer populations on the Pinedale Mesa plummeted by 60 percent.[34] The development is constricting the route of "the largest migrating herd of pronghorns in North America."[35] The region's pronghorn herds summer in the Gros Ventre Mountains and Grand Teton National Park; the park herd is believed to have declined by 60 percent since 1995.[36] The animals pass through the Pinedale Anticline on their way to their wintering grounds in the Red Desert, a vast expanse of drylands east of Green River that stretches south and west toward Utah and Colorado. The pronghorns depend on the Red Desert for winter habitat, and many Wyomingites treasure the fossil-rich area for its archaeological history and its extraordinary landscape.[37] But the Red Desert also harbors rich energy reserves, and in June 2006, the BLM approved a plan that will permit seismic surveys for oil and gas drilling in the area. The plan could allow close to nine thousand gas wells in the region over the next four decades. Some ranchers worry the development may compromise their grazing rights; it could also damage or destroy areas revered by Native Americans, spectacular scenery, and a portion of the Oregon Trail.[38]

In 1885, a young Owen Wister journeyed west to recuperate from a lengthy illness. "The air is better than all other air," the future novelist wrote to his mother as he traveled to a ranch along the Wind River Range, a landscape that the painter Albert Bierstadt had made famous back east with his 1863 canvas *The Rocky Mountains, Lander's Peak.*[39] But the big skies above the Wind River Range and much of the West are increasingly hazy.[40] Sublette County residents sometimes complain of a "brown cloud" hovering over the region.[41] Trucks driving to and from the gas, oil, and coal fields raise clouds of dirt and spew diesel fumes. In early 2006, the Environmental Protection Agency noted two spikes in ozone levels in the Pinedale region; neither spike could be attributed to weather and both appeared to be caused by activity on the ground. "In 2007, the state issued the first ozone alert for southwest Wyoming," reported the *Casper Star-Tribune*.[42] Additional ozone alerts followed—including five in 2008.[43] "When the ozone alerts came, the [Department of Environmental Quality] advised us not to go outside and recreate," Linda Baker, a Pinedale activist, observed in 2008. "And we're in one of the outdoor recreation capitals of the world."[44]

Expanded development in Jonah Field and the Pinedale Anticline could mean even smoggier days in the Upper Green River Valley, with the haze compromising public health and clouding Grand Teton and Yellowstone National Parks and five wilderness areas.[45] One area Forest Service hydrologist believes the air pollution has affected the region's high mountain lakes. Al Galbraith, who monitored "the chemistry of snow and rain falling in the Wind River Range" before retiring from the Forest Service, told the *Casper Star-Tribune* in 2008 that "acidification in high mountain lakes" in the Wind River Range coincided with drilling in the Pinedale Anticline and Jonah Field. Galbraith reasoned that expanded drilling would likely exacerbate the problem, possibly even "lead[ing] to the loss of these fisheries."[46]

Others worry that gas drilling in Sublette County and elsewhere in Wyoming will contaminate the state's aquifers, if it has not already done so. Gas stores beneath the Pinedale Anticline and Jonah Field, locked within the region's geologic formations, went untapped until the early twenty-first century, when combined technological advances in drilling made gas extraction feasible. Hydraulic fracturing, or fracking, is a controversial drilling method that renders gas-rich rock into pay dirt. Fracking involves "pumping sand and fluids—often diesel fuel—into natural gas-bearing rock" under very high pressure, splintering the rock and releasing the gas.[47] The Interstate Oil and Gas Commission stated in 2009 that "there are no known cases of groundwater contamination associated with hydraulic fracturing." Yet at least "89 water wells (most of them industrial) in Sublette County are . . . contaminated" with industrial chemicals. Threats to Wyoming's aquifers raised such concern that the state engineer's office in early 2010 proposed more stringent regulations on "all new residential, industrial, and municipal water wells."[48]

Wyoming Energy Stories

Natural gas extraction is only one chapter of Wyoming's collective energy stories. East of Pinedale, the Powder River Basin provides nearly "40 percent of all coal mined in the United States."[49] The Powder River Basin also yields natural gas pumped from coal-bed methane wells. Dustin Bleizeffer, the *Casper Star-Tribune*'s energy tracker, reported dust "events"—a term that our media apply to terrorist attacks and that the Department of Energy uses to describe the detonations of nuclear bombs—in the Powder River Basin that exceeded acceptable standards over the last few years.[50] The region was traditionally rich habitat for thousands of sage grouse, but grouse populations have declined by 84 percent since the BLM agreed to the "drilling of 51,000 coal-bed methane wells in the Powder River Basin several years ago."[51]

Intensified coal-bed methane extraction in Wyoming's Tongue and Powder River Basins has ignited a feud between the Cowboy State and its neighbor, Montana, which maintains higher water-quality standards. Coal-bed methane extraction requires large amounts of water; the water becomes salty during processing, and some of it is reusable, but some is not. Downstream Montana farmers fear that their crops will be damaged by salinated water discharged into the Tongue and Powder Rivers, both of which feed into the Yellowstone River—yes, the same Yellowstone River that runs through our first national park.[52] Since "thousands more wells are planned" in both Montana and Wyoming, Montana governor Brian Schweitzer aims to keep the Treasure State's water quality high enough to protect southeastern Montana's agricultural fields.[53]

The *Casper Star-Tribune* has called Wyoming "the epicenter of coal-bed methane development in the United States."[54] But Stephen Jackson, a professor of botany and director of the University of Wyoming's program in ecology, calls his home state "the dead center . . . [or] a poster child for some of the [environmental] trends" that scientists attribute to global warming, such as longer and drier fire seasons.[55] The *Star-Tribune* reported that "January and February [of 2006] saw snowpack levels near 130% of average, but by March, that number dropped to 117% and by May 1 after a warm spring, that dropped to 76%." Historically, cool springs meant snowpack kept Wyoming's high ground moist into mid-summer. But warmer temperatures mean snow is melting earlier, and early spring runoff, feeding into rivers and reservoirs, leaves the ground dry and vulnerable to fire. Jackson argues that research linking global warming to more severe fire seasons may well be "the grizzly bear banging on the cabin door."[56]

The Virtual Sublime

Responding to a 2006 presidential "directive to draw upon the best scientific research to address the issue of global climate change," the Department of Energy—the present-day incarnation of the same agency that gave us what Rebecca Solnit calls the "postmodern sublime" in the form of the nuclear bomb—has employed what I call the *virtual sublime*.[57] Edmund Burke's eighteenth-century sublime located the gorgeous and terrible power of God in the fearsome beauty of the landscape. The Hudson River school painters of the transcendentalist era depicted the sublime in storms, cliffs, high mountain peaks, oceans, and dramatic waterfalls, but also located the presence of the creator in quieter landscapes, infused with the balance of the beautiful. The post–Civil War sublime, often laden with Manifest Destiny ideology of divinely ordained expansionism, revealed God's power in the dramatic landscapes of the West, and eventually in the imprint of empire on those landscapes. The technological sublime revealed the power of humans—who harnessed the power of nature through engineering—in the form of romanticized representations of railroads and factories. In the case of Chicago's White City, the electrical sublime indicated a literally brilliant future enabled by technological subjugation of nature. The communications theorist Vincent Mosco contends that widespread adoption of computer technology ushered in the digital sublime, which is "sustained by the collective belief that cyberspace [opens] a new world by transcending what we once knew

Figure 5: *FutureGen*. Image courtesy of the U.S. Department of Energy.

about time, space, and economics."[58] The literary critic Lee Rozelle's ecosublime describes the "awe and terror" one might experience "in the face of a global breakdown."[59]

The virtual sublime offers a virtual reinvention of the world, an Eden with engines displayed on a screen but existing in virtual space. The virtual sublime locates technology in mythic space, where energy is pure and human power is absolute. Human engineering exerts absolute control over all technology—over all of nature—and technological advances solve all problems without creating any new ones. Forget about the terror. Revel, instead, in awe of technology and the people who developed it or who represent it. In the virtual sublime, digital dreamworlds meet nineteenth-century American landscape aesthetics, where together they frame vistas of progress over domesticated and romanticized landscapes. They also meet twenty-first-century American public policy in a collaborative project the DOE calls "FutureGen—Tomorrow's Pollution-Free Power Plant."[60] According to the DOE, FutureGen "will establish the technical and economic feasibility of producing electricity from coal (the lowest cost and most abundant domestic energy resource) while capturing and sequestering the carbon dioxide generated in the process."[61] Digital images of

FutureGen (fig. 5) show a pristine industrial park set amid a Hudson River school–style landscape. The foreground of one FutureGen dreamscape features a building that looks like a postmillennial version of a structure from the White City—a gleaming box of glass panes set atop planes of pale gray concrete, housing the engines of the sublime.

Like the technological sublime, the virtual sublime implies a Godlike human control over nature. Like the White City, FutureGen promises to illuminate and empower America's future. The industrial plant in the FutureGen dreamscape is surrounded by the verdure of a parklike, pom-pom forest where all the trees appear to be identical and the forest's understory is a lawn. Distant mountains recede beneath a cloud-flecked sky and the FutureGen plant's likeness gleams back from a body of perfectly still water. The placid water is a trope right out of the Hudson River school paintings. In his "Essay on American Scenery" Thomas Cole asserted that any landscape representation that did not include water was "defective." The surface of a still body of water, he maintained, "contributes greatly to the beauty of the landscape; for the reflections of surrounding objects, trees, mountains, sky, are most perfect in the clearest water; and the most perfect is the most beautiful."[62] The "still body of water" in the FutureGen visual is a far more appealing image than, say, a mountain of coal or a slag heap. The FutureGen water looks peaceful, but when we see a power plant combined with a body of water, we know the water is probably corralled behind a dam. That still water had better run deep; the FutureGen Industrial Alliance "estimates . . . that the plant will require 2500 gallons of water per minute."[63]

Picturesque, beautiful, and sublime color this vision of FutureGen, but the digitized homogeneity of the landscape indicates subjugation of nature and serves as a metaphor for the loss of biodiversity. With its Hudson River-esque layout, the FutureGen image evokes the progressive, God-blessed narratives of the environment as an aestheticized resource-base that Americans have been conditioned to find in landscape representations since the nineteenth century.

Gutzon Borglum, the sculptor of Mount Rushmore National Memorial, elevated and fused the human to the sublime by placing Presidents Washington, Jefferson, Lincoln, and Theodore Roosevelt shoulder-to-shoulder with the Almighty, whose power, according to Euro-American landscape aesthetics, was embodied in the mountains of the Black Hills. During the presidency of George W. Bush, the DOE offered another rendition of the virtual sublime that synthesized the Rushmore sublime and the electrical sublime. This DOE visual superimposed the head of President Bush on a landscape lush with lawns and framed by mountains. The size of the president's head far exceeded the scale of the Mount Rushmore busts. Godlike, the president's head dominated the left foreground of the frame and floated above a finely trimmed lawn. Beside the president's head, but set at the midground, gleamed a city of power plants where cooling towers and two obelisks dwarfed the background mountains.[64]

We can trace the American passion for lawns back to eighteenth-century England, with Edmund Burke's descriptions of the beautiful.[65] In *The Lawn: A History of an American Obsession*, Virginia Jenkins tells us that Americans "adopted" the lawn "because it fit well with the pastoral ideal of America as a garden"; she adds that lawns are emblematic of human domination of the environment.[66] Beatriz Colomina, an architectural historian, observes that in the United States, the lawn has become a cultural symbol of both democracy and patriotism.[67] The architectural theorist Mark Wigley finds that well-kept lawns are so closely tied to "cultural norms" that "the entire social order is seen to rest on neatly trimmed blades of grass."[68] The power plants in this image were a contemporary version of the visions of American progress through energy technology represented in the 1893 Columbian Exposition. Frederick Jackson Turner spoke of the frontier from Chicago's White City; the president seemed to speak from the frontier of future energy. By casting the president in sublime dimensions over a swath of grass and in front of a twenty-first-century White City, the DOE evoked national ideologies of a God-blessed, pastoral democracy of patriotic citizens whose commander-in-chief exercised absolute authority over technology and all of nature.

Manifest Destiny landscapes depicted the glory rather than the gore and environmental damage of empire-building. FutureGen is supposed to be "pollution-free," and the DOE's virtual sublime depicts visions of future power as if we could produce energy without incurring any environmental costs—not even by mining all that coal.

In July of 2006, the FutureGen Industrial Alliance announced that the FutureGen plant would not be located in Wyoming, which had lobbied hard for the project, and in December 2007, the alliance selected Illinois.[69] However, a November 2009 headline in the *Casper Star-Tribune* announced that Wyoming is "among carbon sequestration pioneers," and at this writing, the state has several carbon sequestration projects in the planning phase.[70] With FutureGen in the Prairie State and carbon sequestration projects powering up in the Cowboy State, we are hitching our wagon to the coal star.

"Clean Coal": An Oxymoron

According to the DOE, "coal is the workhorse of the nation's electric power industry, supplying more than half the electricity consumed by Americans," and "coal-fired electric generating plants are the cornerstone of America's central power system."[71] Coal is also plentiful; the George W. Bush White House estimated that coal stocks would last at least 250 years.[72] FutureGen will employ integrated gasification combined cycle technology—what the DOE calls "clean coal" technology.[73] But "clean coal" is something of a misnomer, as it sounds far more innocent than it actually is. When attached to any type of energy process or project, the term "clean" works like rhetorical bleach; it whitewashes the violent processes of resource extraction and the multiple chains of environmental problems that they trigger.[74] While the system would winnow out many of the particulates that contribute to air pollution, the vanilla label "clean coal" masks one important point: the process generates a great deal of carbon dioxide.

The environmental scientist Tim Flannery records that about "56 percent of all the CO_2 that humans have liberated by burning fossil fuel is still aloft, which is the cause—directly and indirectly—of around 80 percent of all global warming."[75] In a best-case scenario, the carbon dioxide generated by "clean coal" sequencing is collected and stored underground via a process called "capture and sequestration."[76] "There are a lot of parallels between ['clean coal' technology] and nuclear energy," Mark Morey, of Cambridge Energy Resource

Associates, asserted in 2004. "The plants are really expensive to build and there's an issue about disposing of large amounts of [carbon dioxide] waste that could get really costly."[77]

Some oil producers in Colorado and Wyoming have found an industrial use for the carbon dioxide that their current operations generate; they pipe it to their oil fields and pump it into their wells. The gas "'sweeps' additional barrels of oil that otherwise are left in the ground through conventional recovery methods," the *Casper Star-Tribune's* Dustin Bleizeffer explains. The system is so effective, Bleizeffer notes, that some of the companies using it have seen older oil wells "returned to flows not seen since the 1970s."[78] Thanks to carbon storage, Wyoming's Natrona County Salt Creek oil field yielded 3.2 million barrels of oil in 2008, making it Wyoming's third most productive oil field, surpassed only by Jonah Field and the Pinedale Anticline.[79]

Some industrialists assert that if Wyoming's coal plants could capture their carbon dioxide—a procedure that would run the average coal plant about $31 million per year—they could then sell it and pipe it to customers in oil fields, who would use it to boost oil production.[80] Bleizeffer notes that the system would store the carbon dioxide underground, so that it would not contribute to global warming.[81] Assuming that energy companies could successfully "sequester" the gas forever, what progress are we actually making if we store carbon dioxide by using it to pump still more oil, the consumption of which produces even more carbon dioxide? And if a "clean coal" plant or CO_2 pipeline should leak carbon dioxide, "then you're right back where you started, plus you've wasted all that money," warns Dave Hamilton, director of the Sierra Club's Global Warming and Energy Programs.[82]

Others see yet another market for the stuff; carbon credits could be traded on the Chicago Climate Exchange, whose members voluntarily agree to progressively reduce carbon dioxide emissions. Call it "carbonomics"—a manipulation of numbers that produces a measurable change on paper but not necessarily in the atmosphere. "When it comes to the environment, big American companies like to appear green," reports the *Economist,* and members of the Chicago Climate Exchange initiated their program in anticipation of future regulations.[83] If the state of Wyoming were to join the exchange, its businesses unable to meet their emission reduction quotas could purchase carbon credits from participants who exceed their own quotas. Since Wyoming's agricultural lands are considered a carbon sink, the Cowboy State's agricultural producers could sell carbon credits to their industrial neighbors.[84] Grant Stumbough, of the Wyoming Department of Agriculture, "estimates that Wyoming's agriculture industry could reap $7.8 million annually by selling carbon credits."[85] By paying agricultural producers to remain pastoral, big carbon emitters could appear greener even while continuing to discharge carbon dioxide. It is unclear how much—if at all—a trading program might reduce Wyoming's overall emissions of carbon dioxide. Since the United States generates approximately 25 percent of all greenhouse gas emissions, argues Richard Sandor, creator of the Chicago Climate Exchange, "even a small percentage of change becomes significant."[86] But traded carbon credits that "do nothing to diminish climate change," notes Tim Flannery, "are known as 'hot air.'"[87]

New West Dramas

"The West has been raided more often than settled, and raiders move on when they have got what they came for," Wallace Stegner remarked in "The Sense of Place."[88] That is a familiar plotline, but the story is not quite that simple, and not all of the news coming out of Wyoming is so controversial. Some Pinedale landowners have profited by breaking up their property and selling it in parcels to accommodate the influx of new residents. By one estimate, "336,000 acres of ranch land [in Sublette County] may be converted to residential use" by 2020.[89] The state's treasuries are fat with energy industry tax revenues. Thanks to tax dollars and the support of energy companies, Pinedale's elementary school classrooms are outfitted with sophisticated computer equipment and the town's teachers "are the highest paid in Wyoming."[90]

Boom times also meant a spike in jobs. Analysts predicted in 2006 that "about 60,000 highly skilled oil and gas production technicians [would] be needed to meet the nation's energy needs until the end of the decade," and that thousands of those technicians could expect to find work in western Wyoming. Wyoming's then-governor Dave Freudenthal, faced with the possibility of a workforce dominated by transient laborers and hoping to avoid a modern-day version of an Old West drama, said he would prefer that new jobs go to Cowboy State residents, and he hoped that workers from elsewhere who followed the job trail to Wyoming would make the state their home.[91]

Wyoming jobs surged 5.3 percent in 2006—faster than any other state—and the Cowboy State's natural resources and mining jobs jumped 7 percent between November 2007 and November 2008.[92] But demand for energy fizzled along with the American economy. Thomas Power, an economist at the University of Montana, reports that "prices of oil, natural gas, and coal . . . tumble[d] dramatically, by 40 to 70 percent between summer of 2008 and the summer and fall of 2009."[93] Wyoming lost 800 energy jobs in February 2009 alone.[94] Ernie Goss, director of the Goss Institute for Economic Research, found that by the end of 2009, the state had "lost more than 20 percent of its mining employment, or 6,000 jobs, due to the global recession and a significant improvement in productivity."[95] Energy policy experts, however, expect energy consumption to track the economy's progress.[96] Perhaps the arc of a boom looks more like a switchback, as energy consumption follows the trail of the economy.

Urban Frontiers

Most of us who live elsewhere don't see Wyoming or other states when we turn on the lights or start up our cars. I know I don't, even though a portion of the electricity that illuminated my home in Reno was powered by Wyoming coal, and the BLM reports that the electricity for 20 percent of U.S. households and businesses "is produced from coal mined in Wyoming."[97] Contemporary visions of Wyoming's landscape, informed by the myths of the past, show up in popular culture in films like *Brokeback Mountain* and *Open Range*, even though neither of those films was actually shot in the United States. Nor do most of us necessarily realize how much power the collective myths about the

West still exert in American culture. One recent winter, while visiting friends in New York, I walked along Broadway on Manhattan's Upper West Side, noting how my old neighborhood had changed in the years since I had left. I came upon a palatial residential building called The Montana, a name that struck me as ironic. During the years I lived in Montana, the state ranked fiftieth in per capita income, but this place in New York was designed for people with serious money. Were the building's owner's marketing western mythology?[98] This was the *West* Side of Manhattan.

And where were all the Chase Bank branches? They used to be as ubiquitous as Starbucks, so that every few blocks, New Yorkers could get the two things they need most: cash and caffeine. Many of the Chase branches had been replaced by others; one had become a Capital One. It was evening and the bank was closed, but the plate glass window revealed a floor-to-ceiling mural featuring a black-and-white photograph of a rodeo cowboy and bucking horse. Had the pair taken a wrong turn somewhere east of Laramie? The only horses in New York City are pulling tourists around Central Park in hansom cabs, or are living in stables furnished by Manhattan's equestrian elite, or are working for the cops. And the only cowboys are tourists. Bank customers probably don't know that most cowboys are, as William Kittredge describes, a busted-up lot who scrape through a hardscrabble life.[99] Both the bank and the developers of The Montana were likely trading on myths about the West, possibly even more appealing to twenty-first-century easterners—for whom travel is relatively easy as long as they are able to finance it—than they were to the emigrants who risked so much in undertaking their westward journeys.

Front, Frontier, and the Price of Freedom

Perhaps we need those myths about a heroic heritage and a wide open frontier and the freedom it connotes more than ever since the terrorist attacks of 2001 and subsequent American engagement in wars, the fronts of which are not easily determined. "There is nothing like a war to concentrate the minds of citizens on the meaning of patriotism, national identity, and political obligation," observes the historian Robert Westbrook in his study of American patriotism and World War II.[100] In 1941, President Franklin D. Roosevelt rallied the nation when, in an address to Congress, he spoke of the rights of Americans to Four Freedoms: Freedom of Speech, Freedom of Worship, Freedom from Want, and Freedom from Fear. Norman Rockwell's paintings of the Four Freedoms created following Roosevelt's speech were reproduced and widely circulated in the *Saturday Evening Post* and subsequently adopted as posters encouraging the purchase of war bonds.[101] Evocative of a secure, comfortable lifestyle supported by democracy and capitalism, Roosevelt's words and Rockwell's images served as touchstones of American patriotism. The posters were eloquent ideological tools, and the command to "Buy War Bonds" linked American hearts and pockets to the war effort.

"When our country is at war and we ask 'why we fight,'" Westbrook adds, "we are seeking not only an understanding of the aims of a particular war but also a deeper and more general awareness of just what binds us to (or alienates us from) our nation state."[102] The posters indicate that Americans were fighting for these Four Freedoms during World War II, but the generic answer to why the United States fights any war is that the United States fights for freedom. In 2004, the Smithsonian Institution's National Museum of American History opened a permanent exhibit called *The Price of Freedom: Americans at War*, in which American engagement in warfare, beginning with the Revolution and extending to the wars in Iraq and Afghanistan, is rhetorically linked to freedom. Consider the names of the most recent American engagements: Operation Enduring Freedom in Afghanistan and Operation Iraqi Freedom in Iraq. And in recent years, American news media have reported a "showdown in Fallujah" to describe a battle in Iraq and, regarding talks with Iran about nuclear power, a "nuclear showdown."[103] While it is true that news writers must fit their stories into narrow slots of print space or air time, this Old West rhetoric reduces extremely complicated issues, involving many parties, to simplified, polarized binaries, with the United States cast in the role of the frontier sheriff defending freedom.

The New Four Freedoms

"The language of patriotism has been used over the centuries to strengthen or invoke love of the political institutions and the way of life that sustain the common liberty of a people," writes the political theorist Maurizio Viroli.[104] But what is the meaning of "way of life"? What is "liberty"? A few vocal strains of cultural discourse indicate that both are linked to American driving habits. Consider the FreedomCAR. What began in 2002 as a cooperative program between the DOE and the automotive industry is now the FreedomCAR and Fuel Partnership, a coalition that includes the DOE and representatives of automotive manufacturers, energy producers, and utility companies.[105] According to the DOE, the program's goal is "a clean and sustainable transportation energy future that reduces the nation's dependence on foreign oil and minimizes regulated emissions and CO_2, yet preserves freedom of mobility and vehicle choice for consumers." Evidently, domestic—but not foreign—oil will be okay as the program moves toward its final goal of a "Zero Petroleum and Emission" vehicle.[106] The FreedomCAR and Fuel Partnership maintains that "the 'Freedom' principle is framed by: Freedom from dependence on imported oil; Freedom from pollutant emissions; Freedom for Americans to choose the kind of vehicle they want to drive, and to drive where they want, when they want; and Freedom to obtain fuel affordably and conveniently."[107] Such partnerships give me hope for the future. But when people assert that our troops overseas are "defending our freedoms," is *this* what they mean?

By casting the FreedomCAR program in the frame of FDR's Four Freedoms, the FreedomCAR and Fuel Partnership appears to be attempting to link reduced oil imports and increased domestic oil production to the legacy of World War II and to freedom of speech, freedom of worship, freedom from want, and freedom from fear. That is a great deal of ideological freight for any motor vehicle to carry, no matter how cutting-edge the technology.

But American motor vehicles have long been designed to haul the dreams of the nation. Consider, for example, the Jeep. Jeep marketers, who gave us the Jeep Liberty, launched the 2007 "Jeep Compass with Freedom Drive I four-wheel drive." If that isn't enough to convince us that we are pioneers and that freedom is a truck, we can purchase the Jeep Patriot. "Legendary capability. Bold new territory," DaimlerChrysler announced on the occasion of the Patriot's 2007 launch. "The name says it all: Patriot. Proud, authentic, resourceful, capable."[108] The Jeep Patriot draws on its "heroic heritage," the automaker added, and equipped with Freedom Drive I or Freedom Drive II, the vehicle would provide "fun, freedom, [and] utility." DaimlerChrysler insisted that the Jeep Patriot was "conquering new territory."[109] With its promise of a fuel-efficient vehicle, DaimlerChrysler may have set out to "conquer new territory," but it did so with old rhetoric. "At the core of the Myth [of the Frontier]," asserts Richard Slotkin, "is the belief that economic, moral, and spiritual progress are achieved by the heroic foray of civilized society into the virgin wilderness, and by the conquest and subjugation of wild nature and savage mankind."[110] Makers of sport-utility vehicles have traded on this heritage for years, and certainly some of the vehicles are about as big as the covered wagons the emigrants drove across the country. Think about the names of a few light trucks: Chevy Blazer, Tracker, Trailblazer, and Avalanche; Dodge Durango; Cadillac Escalade; Ford Excursion and Expedition; GMC Yukon Denali; Isuzu Rodeo and Trooper; Land Rover Discovery; Range Rover; Lincoln Navigator; Mercury Mountaineer; Subaru Forester; Toyota Highlander and Sequoia; Volvo Cross Country; Hyundai Santa Fe. Nissan's Pathfinder bears the national nickname of explorer John Charles Frémont. "Ads that show people 'conquering' natural elements are expressing me-first anthropocentrism," observes the communications theorist Julia Corbett.[111] We might say the same about the implied drivers in those images that feature only the machine.

Many SUV ads of the late twentieth and early twenty-first centuries have been contemporary versions of expansionist-era landscape paintings. Crafted according to Manifest Destiny aesthetics, they depict sublime landscapes that are to be admired and then domesticated by the technology of empire embodied in the SUV. Such images became so common in the years surrounding the bicentennial of the Lewis and Clark expedition that they were nearly indistinguishable from one another. Preliminary images of the Jeep Patriot depicted the vehicle in western canyon country, directly linking it to a heritage of American landscape mythology.

The automaker ascribed to the Jeep Patriot, and presumably to its drivers, the characteristics that Frederick Jackson Turner attributed to frontier explorers and settlers. The belligerent phrase, "conquering new territory," projects all the violence of Manifest Destiny ideology—with the empire-building, genocide, and environmental damage that informed it. It hardly seems coincidental that the Jeep was originally a military vehicle, that the Jeep Patriot shares its name with both an antiballistic missile and a day dedicated to the remembrance of the victims of the 2001 terrorist attacks, or that the Hummer, also originally a military vehicle, was marketed as a haute rig before its American demise. Nissan, too, joined the automakers' war games with the 2006 introduction of the Nissan Armada; the Armada's drivers can ride with the assurance that they are protected by a metaphorical fleet.

The Jeep Patriot is an eloquent token of that hybridization of earlier lines of patriotism: as Jonathan Hansen puts it, the equation of "freedom with acquisition and liberty with property" in the nineteenth century, and as John Bodnar notes, the "veneration" of male warriors and high-tech warfare in the twentieth century.[112] Powered by cold war and frontier rhetoric, the FreedomCar program and the marketers of the SUV as personal tank or latter-day covered wagon celebrate technology wedded with militarism and patriotism and blur the distinctions between safety, material goods, and the freedoms protected by the Constitution. As Terry Tempest Williams argues in *The Open Space of Democracy*, "We have made the mistake of confusing democracy with capitalism and have mistaken political engagement with a political machinery."[113] Take, for example, the popularity of the Fox News financial news program *The Cost of Freedom* and Craig Frazier's poster of an American flag in the shape of a shopping bag, designed for the "America: Open for Business Campaign" following the 2001 terrorist attacks. Such cultural artifacts reflect a vein of patriotism that casts consumption as both a patriotic gesture and an act of national defense, as if we were a nation of militant shoppers.

We are not a nation of militant shoppers. But one thing this nation does consume more than any other except China, which has more than triple the population of the United States, is energy—not just for motor vehicles and businesses but also for bigger homes.[114] According to the *Wall Street Journal*, "the average American [burns] five times as much energy annually as the average Chinese citizen."[115] The Department of Energy reports that in 2009, Americans burned through 22.8 trillion cubic feet of natural gas and nearly 1 billion tons of coal. We also consumed 18.7 million barrels of crude oil and liquid fuels per day in 2009, which was, by all accounts, a lean year.[116] By contrast, the International Energy Agency reports that 1.5 billion people worldwide currently live without access to electricity.[117]

The Northwestern Energy Frontier

In *The Fatal Environment*, Richard Slotkin writes that pre–Civil War frontier dreams were fueled by "the belief that westward expansion could provide an inexhaustible stimulus to economic growth."[118] Our media reflect those "light out for the territories" dreams. In 2007, for example, an executive for Royal Dutch Shell described Alaska's Beaufort Sea as "a great frontier," and invited the media aboard a drilling rig named the "Frontier Discoverer."[119]

In a 2006 issue of *BusinessWeek*, the Houston-based True North Energy Corp, attempting to attract investors, ran a full-page ad with a headline that read: "Alaska! Focus on America's Last Emerging Energy Frontier." The copy is not surprising; one of Alaska's state nicknames is "the Last Frontier." Ad graphics include a snowscape featuring a gas drilling rig illuminated by the sun—the technological sublime mounted in one of the most myth-laden and forbidding of American landscapes. The company's webpages feature stunning photos of the technological sublime: a drilling rig, backlit by the sun, surrounded by snowfields; another snowscape, with ice and snow mounds in the foreground lined up with a drilling rig and buildings in the background, smoke billowing like clouds across the horizon; a shot of shimmering Anchorage embraced by mountains awash in shadow and twilight. The images are spectacular, and they are what I expect.

What I don't expect are the polar bears. Two polar bears march across the ice on the company's page describing its "Competitive Landscape." On the page listing True North Energy's "Management Team," six polar bears stand, sit, or recline on a gleaming ice sheet.[120] Sea ice provides critical habitat for polar bears, but with rising temperatures, the ice sheets are melting, and in 2007, the U.S. Geological Survey projected that "two-thirds of the polar bear's habitat would disappear by 2050." In 2008, invoking the Endangered Species Act, the Department of Interior classified polar bears as a "threatened" species but simultaneously extended protections to "offshore oil and gas drilling [operations] in prime polar bear habitat off Alaska's north coast."[121] The Management Team bears look up at the camera, with the names of True North Energy's executives superimposed across the screen. Perhaps whoever designed or approved these pages did not know, or did not care, that polar bears are in trouble. Or perhaps the image is supposed to demonstrate that any polar bears in the vicinity of True North Energy's operations are doing just fine. Or maybe this is the work of a graphic designer who wanted to give the polar bears a voice in the story of energy consumption.

The ad and the webpages are chapters in a twenty-first-century western—a familiar story in different media and with different characters. And the western, writes William Kittredge, is "a story inhabited by mythology about power and the social utility of violence, an American version of an ancient dream of warrior righteousness."[122] I would add that violence against the environment is also violence against human communities. In Alaska, the "frontier" that True North Energy would have investors imagine is receding as the earth's temperature climbs. The Inupiat coastal village of Shishmaref is relocating inland because the permafrost is melting out from underneath it, and its coastline, stripped of its protective belt of ice, is eroding into the sea.[123] Other Alaskan villages are following, but such problems are not unique to the United States. The Environmental Justice Foundation estimates that "around the world, as many as 150 million people may become 'climate refugees' because of global warming."[124] As the western writer and political theorist Linda Hogan observes, what happens to the land is what happens to people.[125]

Big Open (for Business)

Meanwhile, back on the range, the language of the expansionist era runs throughout Wyoming's energy projects. Like Thomas Moran's romanticized representations of Green River, which replaced industrialism with a Euro-American vision of landscape and Natives, much of the narrative of American energy remains burnished with the rhetoric and myths of the frontier. Depictions of industrialism often glorify the technological or the virtual sublime and spin the story into a teleological narrative.

The names of the projects imply that both the energy producers and our government hope to trade on some of the flashier themes—freedom, independence, divinely ordained mission, terrain conquered by technology—from the expansionist chapters of the nation's creation stories. Gas pumped from Jonah Field is siphoned into the Pioneer Gas Plant.[126] The governors of California, Nevada, Wyoming, and Utah are backing development of the Frontier Line, an improved electrical system that will be powered by Great Basin reserves, including fossil fuels, "clean coal," and renewable sources.[127] And as the Pony Express delivered news of the world to the western regions from 1860 through 1861, so the Rockies Express Pipeline transmits natural gas eastward. Nicknamed REX by its sponsors, the $6.8 billion pipeline extends more than 1,600 miles from a hub in Colorado into Ohio. The REX pipeline system can siphon about 1.8 billion cubic feet of natural gas—primarily from Colorado and Wyoming, although perhaps that will change with additional Rocky Mountain extraction projects pending—every day to customers in the Midwest and East.[128]

Pinedale may be emblematic of the good and bad that happen when big business rolls into (some might say over) a small town, but similar stories are in progress—or soon will be—all over the intermountain West. In 2005, the BLM received more than 8,000 Rocky Mountain-region drilling permit applications, nearly double the number the agency fielded in 2002.[129] According to a report in the *Casper Star-Tribune*, natural gas production alone in the "Rockies is projected to double in the next 20 years, surpassing production in the Gulf of

Mexico."[130] Questar, one of the Pinedale Anticline's big operators, intends "to drill 4,000 oil and gas wells in a 150,000-acre area along the Colorado-Wyoming border over the next 30 years."[131] Instead of an American flag, explorer John Charles Frémont might as well have waved an "America: Open for Business" shopping bag poster from the Wind River Peak he climbed in 1843.

"Energy development drives the Wyoming economy and our state provides critical energy resources for the entire nation," Bob Bennett, then state director of Wyoming's BLM, proclaimed in 2006.[132] But Stephen Jackson, an ecologist at the University of Wyoming, takes a longer-term view of his state's economy. Jackson argues that Wyoming should stand "at the forefront of developing new technologies to reduce humans' impact on global warming." Wyoming's "economy is heavily dependent on fossil fuels," he cautions. "But we should be thinking about a transition to a different type of economy. We need to prepare for a post–fossil fuel kind of economy."[133] Accordingly, as of this writing, some Wyoming energy producers are setting up wind farms on the Big Open.[134]

The Rhetoric of Progress, the Aesthetics of Power

All the way across the state from Pinedale, tangled notions about patriotism, war, energy, terrain, commerce, and cowboys touch down on the football field of the University of Wyoming's War Memorial Stadium, which is one of a handful of war memorial stadiums in the United States. "The War," as the locals call it, opened in 1950 and was "dedicated as a 'living memorial' to Wyoming's men and women who served in World War II."[135] Following a $5 million gift from energy industry entrepreneurs John and Mari Ann Martin and Mick and Susie McMurry in 2005, War Memorial Stadium's gridiron was named Jonah Field. "This donation is fueled by Wyoming's rich oil and gas resources," Mick McMurray stated when the gift was announced. "Jonah Field in particular will contribute thousands of jobs and billions of dollars to the citizens of Wyoming during the next decades."[136] Now when UW's Cowboys take to their upgraded field, they play on a symbol not only of twentieth- and twenty-first-century patriotism but of the benefits and costs of energy development in Wyoming's Green River Valley.

Downstream of Pinedale and Green River, energy companies also have their eyes on the shale-rich earth near Flaming Gorge, a popular National Recreation Area that stretches across Wyoming and Utah. John Wesley Powell named the red rock gorge in 1869, when he remarked the sun-splashed canyon walls on his first trip down the Green River.[137] Nearly one hundred years later, Lady Bird Johnson floated the Green on a trip in which she sought to promote conservation of the nation's beautiful landscape. "Enjoy the beauty of your hills and protect it for your children," the First Lady advised onlookers as she presided over the dedication of Flaming Gorge Dam.[138] Such words, paired with a massive dam project, seem ironic now, but they may well have been prescient.

Thomas Moran returned to Green River in 1873 to accompany Powell's third expedition down the Green and Colorado Rivers, and Moran produced more than a hundred works of art from material gathered during this trip.[139] Powell "contradicted notions of the West's fecundity and its ability to support American-style settlement . . . but the landscape was already culturally 'mapped,' its destiny manifest as 'promised' land," writes the eco critic Rick Van Noy. Boosters had described the West as the "Garden of the World," and the belief in an American Canaan was too deeply entrenched to be uprooted by scientific observations. "Historians . . . have concluded that Powell's revolutionary ideas were poorly received because they conflicted with a deep-seated and fervently held optimism about the West, and because their recommendations for commonwealths gathered around watersheds flew in the face of the ethics of capitalism and resource exploitation," argues Van Noy.[140] Lawmakers, however, welcomed Moran's sublime vision of the West, and they purchased his *Chasm of the Colorado* for $10,000.[141] Like Emanuel Gottlieb Leutze's 1861 tribute to Manifest Destiny, *Westward the Course of Empire Makes Its Way*, and Moran's *Grand Canyon of the Yellowstone* (see fig. 4), *Chasm of the Colorado* was hung in the Capitol building, testament to conflicting visions of the landscape as sublime panorama and resource cache, and to the influence of romanticized representations of the American landscape on public policy.

John Wesley Powell and Thomas Moran could not have imagined that 130-odd years after their expedition, uranium mining companies would eye the regions surrounding what had become, in 1919, Grand Canyon National Park, and that the U.S. Geological Survey would undertake studies to determine the potential effects of uranium mining on the park's groundwater, the Colorado River, and regional aquifers.[142] Nor could Powell and Moran have envisioned nuclear power plants and what industry is calling the "nuclear renaissance."[143] The Department of Energy and the Obama White House call nuclear power "clean and safe," and nuclear power is an important component of what the Obama administration calls "clean energy."[144] Terms like "safe" and "clean" are relative. Nuclear power is only as reliable as the humans who manage and handle it, and the waste it generates is so deadly that it poses threats to public health and national security.

The term "nuclear renaissance" fixes nuclear power in mythic space, where it is free of radiation and mining waste; science and art intertwine, every nuclear power plant is a Sistine Chapel ceiling, and every nuclear reactor harbors sublime sparks of divine energy shed from God's fingertips. Back on the ground, uranium mining generates yellowcake—toxic tailings with a name that sounds like a project for a fifth-grade home economics class. Some of the uranium to fuel the nuclear renaissance is likely to come out of Wyoming, where uranium mines were active until prices took a dive in the 1980s. Two Colorado-based companies have joined forces to reactivate uranium mining in eastern Wyoming. And this is where the rhetoric hits me in the gut: the companies have formed what they call the Sand Creek Joint Venture, and their operations will extend for 92,000 acres, starting from the town of Shawnee.[145] "Where are the Shawnee now?" the poet Mary Oliver wonders in "Tecumseh."[146] The town of Shawnee is named for an eastern tribe who were pushed progressively westward. Sand Creek, Colorado, was the scene of the horrific 1864 massacre of two hundred peaceful Cheyenne by a volunteer army.

All of this is the logical outcome of the nineteenth-century project of Manifest Destiny: the carbon dioxide–spewing coal plant named for a mountain man; the uranium company called after a place once drenched with Cheyenne blood; the FreedomCAR and the Jeep

Patriot; the notion of Alaska as the last energy frontier; the carbon dioxide pocketed underground; the "mythematics" of carbonomics. Even "Wyoming" is a Euro-American version of a word from the Lenape—who were also pushed westward—meaning "great plains."[147]

"We need new stories, new terms and conditions that are relevant to the love of land, a new narrative that would imagine another way, to learn the infinite mystery and movement at work in the world," writes Linda Hogan.[148] William Kittredge hammers a like refrain. "We need to invent a new story for ourselves," he insists, "in which we live in a society that understands killing the natural world as a way of killing each other."[149]

Many contend that natural gas extraction is among the most important chapters of that new story. The energy economist Peter Odell predicts that "natural gas will undoubtedly become the prime energy source by the second quarter of the 21st century."[150] The Environmental Protection Agency states that "the burning of natural gas produces nitrogen oxides and carbon dioxide, but in lower quantities than burning coal or oil."[151] That is why energy producers and politicians call natural gas "clean."

Ambassador Richard H. Jones, deputy executive director of the International Energy Agency, warned in 2009 that "a continuation of current trends in energy use puts the world on track for a rise in temperature up to 6 degrees centigrade and poses serious threats to global security."[152] James Hansen, a climate scientist at NASA and one of the world's top climatologists, asserted in 2006 that if we did not reduce carbon dioxide emissions within ten years, we could pass some "tipping points" in the process of global warming. "If the ice sheets begin to disintegrate . . . [i]t will be a situation that is out of our control," Hansen argued.[153]

The lower carbon content of natural gas is no small benefit. "Gas Could Be the Cavalry in Global Warming Fight," Mark Williams, of the Associated Press, dipping into a well of frontier rhetoric, reported in 2009.[154] But natural gas alone will not save us, and the energy expert Daniel Yergin cautions that "diversification [is] the fundamental starting principle of energy security."[155] Nevertheless, the International Energy Agency predicts that natural gas will "continue to play a bridging role in meeting the world's sustainable energy needs."[156]

Yergin and Robert Ineson call the combined advances in drilling technologies that have liberated natural gas in places such as the Pinedale Anticline and Jonah Field a "natural gas revolution." They claim that "the U.S. has about 90 years of proven and potential supply," a number they believe is likely to rise as geologists identify additional stores.[157] Others add that technological advances in drilling will release huge supplies of natural gas globally.[158]

The "one serious obstacle to development of shale resources" in the United States, Yergin and Ineson warn, is water. But, they contend, because most gas stores lie well below aquifers, "hydraulic fracturing of gas shales is unlikely to contaminate drinking water." Nevertheless, they caution that "the risks of contamination from surface wastes, common to all industrial processes, requires continued care."[159] In drylands such as Wyoming and much of the American West, where rain never did follow the plow, summers have been growing longer, hotter, and drier. In lush areas like the northeastern communities above the Marcellus Shale, millions of people depend on limited water resources. Although we put a price on water, its value is inestimable.

Pollster Frank Luntz finds that Americans are, in general, future-oriented. "We Americans perpetually have our feet on the gas pedal," he writes. "We're staring straight ahead through the windshield, never pausing to glance back in the rearview mirror."[160] Counseling against such "historical amnesia," Vincent Mosco cautions that "perhaps the greatest mistake people make about technology is to assume that knowledge of its inner workings can be extrapolated over years to tell us not only where the machine is heading but also where it is taking us."[161] In places such as the Pinedale region, we may well see our destiny manifest.

What lies on our energy horizon? We are working that out on our landscape, in laboratories all over the nation, and at forty-six project sites that the White House and Department of Energy call Energy Frontier Research Centers.[162] When government and industry leaders decide that the timing is right, they will show us pictures of the virtual sublime, the aesthetic of pure and absolute power.

Pinedale

The cultural critic Eric Schlosser calls fast food restaurants "the shock troops" of urban sprawl, and little Pinedale has its share.[163] Snickers bars notwithstanding, I was hungry by the time I reached Pinedale, and after the minute it took to drive through downtown, I doubled back and pulled into the parking lot at Subway, grateful for cheap, predictable food. After lunch, I stopped at the local office of the Forest Service. I wanted to see the industrial landscape I'd read about, and the rangers advised me to go up the adjacent mountain road into the ski areas. From that high ground, they said, I could get a good view of the gas fields.

I drove up. The day was storm-blown, and shafts of sunlight broke through dense clouds and scattered rain. I pulled over and got out of my car but could see nothing below but rain and clouds. I drove higher, pulled over again, and walked through the sagebrush and conifers beside the road, where I found empty soda cans and piles of deer scat. In the valley below, meadows of clouds bordered lakes that shimmered or dulled to slate in the shifting light. I drove higher, up to the snow line, got out again into the cool air with my windbreaker pulled tight around me. Green lichen papered the roadside rocks. Below me, Pinedale's gas fields remained obscured by clouds, as if I were looking at a landscape composed by Thomas Moran, unmarked by industrialism. Ravens gargled just overhead, and I could hear the distant voice of the Wind River. "The landscape is just a situation / of wind breaks and wind permissions," writes the Wyoming poet James Galvin.[164] I stood for awhile looking at the gray faces of the Winds. The mountains were veiled in clouds, and I felt that ache I sometimes feel in the presence of such staggering beauty. I wished I could stay long enough to truly know this place.

Wallace Stegner warned against the dangers of "migratoriness," and he called the wanderer "adventurous, restless, seeking, asocial or antisocial . . . a discarder or transplanter, not a builder or conserver." American culture, he said, lauds "motion" as "a form of virtue," but Stegner saw it as a fear of commitment to place and community coupled with an addiction to change and the hope of "something better."[165] He blamed this transience for the boom and bust cycles that leave their imprint on the land and its communities. It's an unsettledness I know intimately, a Jack Kerouac wet dream of endless opportunities that masks fear and a failure to connect deeply to people and places. Wyoming's landscape buffered the husk of my car, and I marked the miles on my map, gorging on scenery as if open space could fill my own emptiness. I knew my driving wasn't doing the environment any good, and I knew that many people in Pinedale were working hard to protect their homeplace from environmental damage and the boomerang of yet another bust. As I stood looking out at the Winds and breathing the evergreen air, I thought that anyone living here must so love this place. How could they not? Sometimes, when I come close to such places, I wish I could stay and make my life there—give up this crazy marriage to my car.

Conclusion

Green Patriotism:
From "Moral Geography" to "Moral Ecology"

> We have no choice but to live in community.
>
> William Kittredge, *Owning It All*

The works of nineteenth-century American writers and landscape painters often reflected the belief that God was to be found in nature and that "moral values [were] inscribed in the landscape"—what Amy DeRogatis calls a "moral geography."[1] According to Angela Miller, landscape imagery also played an "essential role" in the formation of American nationalism.[2] Post–Civil War American landscape artists frequently conjoined images of the sublime frontier and the notion of building an American empire. Since the sublime was understood as the province of God, such works served as totems that sanctified empire-building and melded the idea of the divine with that of the frontier landscapes of the Far West. W. H. Goetzmann and W. N. Goetzmann observe that Thomas Moran's "*Mountain of the Holy Cross* became an archetypal image of a Christian nation, an outward sign that God himself had blessed the westward course of empire."[3]

When Frederick Jackson Turner spoke at the World's Columbian Exposition of the influence of the frontier on American character, he conjoined the idea of the sublime frontier with the ideals of American democracy. His venue, Chicago's White City, with its spectacularly electrified neoclassical buildings, exemplified the triumphs of the technological sublime. In delivering that particular talk from that particular place, Turner established ideological links between democracy (and nationalism), the sublime (or divine) frontier, and the technological sublime of engineering. Contemporary American landscape aesthetics and the rhetoric that surrounds them, in the employ of commercial or political ends, resonate with nation-building mythology and the promise of the techno-industrial sublime. Such images link the American landscape with patriotism, engineering, and industrial development, repeating the overworked plotline of the Manifest Destiny era.

A twenty-first-century sequel to the old story that began with a search for Eden and visions of a "moral geography" could be based on what some people are calling a "moral ecology," a term defined by a group of scholars in the 1980s as "the web of moral understandings and commitments that tie people together in community."[4] The phrase has been co-opted and reinterpreted since then, but for our purposes, think of "moral ecology" as the awareness that our behavior may profoundly affect people and places nearby and far away, and that such awareness, in turn, may affect our behavior.

"What we need next is a new ethic—call it an 'ecological ethic of care,' call it a 'moral ecology,'" writes Kathleen Dean Moore. "It's an ethic built on caring for people and caring for places, and on the intricate and beautiful ways that love for places and love for people nurture each other and sustain us all."[5] "Humankind," David Orr adds, "is only a small part of a much larger enterprise. Science, too, is a small part of a larger moral ecology aiming to redeem better possibilities than those in prospect."[6] Recognition of a "moral ecology" could revolutionize the way we live. What if we were to think of our economy as a subsystem operating within a set of global ecosystems, rather than as separate from, or in control of, our environment? How might it affect our daily practices? How would such a view influence the ways that we choose to represent, and partition, nature? How might such a mindset influence urban development, agriculture, forest and range management, energy extraction and development, and even national defense?

Upon This Rock, We Build Our Civil Religion

A United States Marine Corps recruitment ad that was launched in 2002 and ran for several years (and as of this writing is still available online) fuses cultural ideas about landscape, scenery, land use, and patriotism. Titled *The Climb*, the commercial opens with a scene of a virtual desert landscape, where clouds roll at high speed across a deep blue sky. We are to imagine that this is somewhere in the American Southwest, and the soundtrack features wind and thunder and then the sort of rattle, flute, and drum sounds that might accompany the "Indian scenes" of a late twentieth- or early twenty-first-century Western. With its flat terrain punctuated by buttes, rock columns, and spires, the image is convincing except for one critical detail: the floor of this virtual desert is drenched in shades of green, as if it were northeastern pastureland.

Then the camera switches from the verdant, virtual desert to real desert, like the red rock drylands featured in ads for the Jeep Patriot and so many SUVs, in which the technological sublime displaces the divine. The camera focuses on a young man, clad in a tank top, khaki pants, and sneakers, who scampers up a rock wall and begins climbing without water or gear. While anyone who can walk can hike at least some terrain, climbing requires technical skill, tremendous strength, and—to borrow a phrase from the military—"uncommon valor." This man climbs. As he advances, the music segues to a military theme. His ascent would shame an American superhero. This guy's power doesn't come by birthright, kryptonite, or spider bite; he is all muscle, skill, and determination. He scales a column of rock, clinging by his fingertips and propelling himself with his toes. His climb is almost entirely vertical, and that's the easy part; at one point, hundreds of feet above a chasm, he dangles by the fingertips of one hand before hauling himself upward.

The camera pulls out and we get a God's-eye view of a sublime desert landscape. We watch the young man—in the tradition of the nineteenth-century sublime, tiny in comparison with his surroundings—climb hundreds of feet up a rock wall. As he works his way toward the top, successive images flash across the wall as if it were a screen. The images, of the same sublime scale as the wall, are also transparent; the rock always shows through them. We are looking at military mythology layered upon landscape mythology.

The shadow of a helicopter hovers above the shadows of a unit of marines running into action. A drill sergeant barks orders, presumably at the young man, who is no bigger than a mouse as he climbs the wall beside his superior. A gentle-eyed marine, in full combat gear, smiles as he holds his rifle. Another marine sits at a table and spoonfeeds a little girl. Another unit, in what could be a colorized film clip of the marines raising the flag at Iwo Jima, hoists an enormous American flag that flutters across the breadth of the rock. When the climber reaches the summit, yet another marine, ghostly, gives him a hand up as a voiceover pronounces, "The passage is intense. But if you complete your journey, you will find your destiny among the world's greatest warriors. The Few. The Proud. The Marines."[7] The climber, surrounded by a desert landscape that appears to be part real, part virtual (but no longer green), is suddenly transformed into full dress uniform. The final scene shows the young man, now one of the Few and Proud, at the head of his squadron; their image is superimposed on the full length of the rock wall.

The landscape doesn't just figure prominently in this brief narrative. The landscape *is* the narrative, as much as the climber and the United States Marine Corps are the narrative. Like the nineteenth-century lawmakers who imagined the West as a garden, and like Nevada's 1894 boosters, who envisioned converting "worthless" desert into farmland by harnessing the Truckee, Carson, and Walker Rivers, the commercial's creators have envisioned their virtual desert through a lens tinted by Manifest Destiny aesthetics. They have framed the landscape as pastoral, picturesque, and beautiful in the opening scene by surrounding the sublime rock forms with a carpet of green. This southwestern desert—except for its rock columns, buttes, and spires—looks like a version of the lush, northeastern meadows that inspired the pastoral paintings of the Hudson River school artists.

The shots of actual desert landscape in the ad are successors to Thomas Moran's paintings of the Grand Canyon, which he made after he accompanied John Wesley Powell's 1873 expedition down the Colorado River. Far from greenwashing the desert, Moran portrayed his sense of the forbidding grandeur of the region. Congress purchased Moran's *Chasm of the Colorado*—one more image in the album of American nationalism—and hung it in the Capitol (even though policymakers would eventually disregard Powell's environmentally sound recommendations for settlements clustered around watersheds). Those gestures—enshrining Moran's aesthetic vision and passing over Powell's conservative water and land management plans—manifest conflicting views of the landscape as scenic panorama and land to be conquered, by settlement or warfare, for its cache of resources. The Marine Corps commercial, created nearly 130 years after Congress hung Moran's *Chasm of the Colorado* in the Capitol, reflects these conflicting views. The opening scene's emerald bottom-lands indicate a bountiful Promised Land, but the rock formations and the soundtrack's wind, thunder, and "Indian scene music" indicate that this spectacular territory must be conquered and cleared of its native inhabitants to access those resources. The commercial travels a nineteenth-century path from its depiction of a green wonderland to footage of the young man conquering forbidding ground.

The desert where the young man climbs represents the most daunting region of the old frontier, and the climber tests himself against the earth. When he reaches the top, he is transformed into a military version of Frederick Jackson Turner's heroic frontiersman. Like the landscape of the Nevada Test Site, the marine's desert serves as a proving ground for military ends. As John Bodnar notes, "The earliest view of a virtuous nation of equals gave way by the late nineteenth century to a dream of a powerful nation rooted in the desires of powerful men and women who supported it for order and moral certainty at home and in the world." The "moral geography" of the early nineteenth century had ceded ground to the "moral certainty" of the late nineteenth century.

We might argue that some of us never left this phase; it was "moral certainty" that gave us the original name of "Operation Infinite Justice" for the American invasion of Afghanistan. "This position, often presenting patriotism as a virtue," Bodnar continues, "was heavily influenced by the more aggressive sentiments of nationalism and the quest for domination of others, both inside and outside the United States. In this version true patriots were often represented as male warriors." This ideal intensified after World War II, he adds, and by the 1980s, visions of this strain of patriotism "offered spectacles of power and the veneration of elite groups of warriors."[8] The voiceover in the commercial echoes this sentiment when it identifies the U.S. Marine Corps as "the few" and "the world's greatest warriors."

The commercial elevates the human to the sublime by superim-

posing human figures on the rock wall at a scale that appears to far
surpass that of the sixty-foot presidential busts at Mount Rushmore,
rendering the U.S. Marines as gods rather than mere humans. About
a truck advertisement that depicts a similarly gargantuan man, Julia
Corbett observes that "such exaggerated domination intentionally
positions humans at the top of a pyramid, instead of belonging equally
to a biotic community."[9] When the Iwo Jima–prototype marines raise
their flag against the rock wall, they fuse national icons: the frontier landscape, the flag, the United States Marine Corps Memorial
in Washington, D.C., and the Iwo Jima Memorial on the grounds of
the Marine Military Academy. The result is an eloquent blend of ideologies identifying patriotism with frontier mythology, the American
landscape, war, and the warrior. The military sublime always includes
the technological sublime; here, both convene in the shadowy forms
of the helicopter and gun-toting unit and in the weapon in the arms of
the smiling marine. The behaviors sanctioned by this merging of ideologies include, but are not limited to, acts of environmental violence
such as the bombing of American deserts, the militaristic approach to
forestry that Gifford Pinchot called "tree butchery," and coal, oil, and
gas extraction—in the West and elsewhere—for the sake of what the
Department of Energy calls "energy security."[10] In the nineteenth-
century tradition, evidence of environmental violence is excluded
from the picture.

"Unlike foreign patriotisms, American patriotism has almost nothing to do with notions of blood and soil," writes Jonathan Foreman
in *The Pocket Book of Patriotism*. "We, alone, are a people dedicated to
a proposition." American patriotism springs from the ideals of our
nation's founders, he adds, and it remains inspired by "faith in human
possibility."[11] Foreman has it partly right. But a few centuries of traditions have demonstrated that American patriotism has *everything* to do
"with notions of blood and soil." Early documents, such as Thomas
Paine's *Common Sense*, with its repeated references to the American
continent and to the violence wrought by the English soldiers on the
colonies, and Abraham Lincoln's "Gettysburg Address," which joins
the nation's creation, the continent, blood, and soil in "a new birth of
freedom," both attest to that. So do our enormous body of landscape
imagery and our frontier and war mythologies, all intertwined and

Figure 6: Chester Arnold, *On Earth As It Is In Heaven*, 1996. Oil on canvas. Collection of Nevada Museum of Art, acquired for the museum by Mr. William Abernathy in memory of Janet Abernathy.

active in contemporary culture. "Notions of blood and soil" show up in settings such as the University of Wyoming's Jonah Field at War Memorial Stadium and in the Marine Corps recruitment ad, which fuses men, war, flag, and landscape.

The Marine Corps commercial echoes the warrior myths of "how the West was won" by imposing military scenes on iconic western terrain and by casting the warrior as both conqueror and heroic guardian of that domain. The commercial is a repetitive chapter in an old story. And as William Kittredge cautions us, "We need a new story, in which we learn to value intimacy. Somebody should give us a history of compassion, which would become a history of forgiveness and caretaking."[12] The U.S. Marine Corps appears to be crafting at least a new angle on an old story by including the gentle-eyed, smiling marine and his colleague who spoonfeeds the little girl. The commercial reflects the military's diversity: several of the marines featured are people of color and one member of the squadron in the final scene is female. Yet the commercial's central figure is a white man, as is the ghostly marine who greets him at the apex of his climb. The rattle, flute, and drum sounds of the opening scene are a trope from Western films and television programs meant to suggest that the climber is in "Indian country."[13] (Richard Slotkin observes that during the Vietnam War, the United States military frequently described enemies as "Indians" and hostile territory as "Indian country.")[14] The commercial replays this metaphor by including the rattle, flute, and drum music and by recording the young man's triumphant climb in the American Southwest—territory that was long-contested by the United States government and several Native American tribes. Expansionist mythology subsumes the stories of the Indian Wars and the commercial's "Indian scene music" quickly yields to triumphant military music. But this is a commercial—designed to sell a set of values. The final image merges the warrior and his squadron with the sublime landscape, redeploying the Myth of the Frontier.

Green Patriotism

"What we need most urgently is a fresh dream of who we are, which will tell us how we should act, stories about taking care of what we've got . . . a story about making use of the place where we live without ruining it," William Kittredge writes in *Taking Care.*[15] Frontier mythology mandated that settlers and the military take the continent by force, reshape it into pastoral terrain, and plumb all the other resources they could find between the coasts. This was the vision of the Eden- and Canaan-seekers and of those with flashier dreams of empire. Many energy companies tap this mythology as they move beyond the coasts in their search for oil beneath the sea and north to "the last frontier" of Alaska, and as they eye the Arctic National Wildlife Refuge. Of the ranch where he was raised, Kittredge declares that realizing the vision of the American pastoral amounted to "warfare against all that was naturally alive in our native home."[16] "Moral geography" gave early settlers the impetus to find God in nature, but it also gave many of them a persistent vision of a lost Paradise and Promised Land, tangled dreams reflected in policies that categorize landscapes as scenic or utilitarian resources.

Kittredge's work presents an alternative vision that is also reflected in the writings of Barbara Kingsolver, Rick Bass, David Orr, Kathleen Dean Moore, Terry Tempest Williams, Sandra Steingraber, and many others, whom the environmental writer Richard Nelson calls "patriots for the American land."[17] We also see this alternative vision in the efforts of scientists such as NASA's outspoken James Hansen and the teams at work at Oregon's Andrews Experimental Forest; activists like the citizen coalitions of Pinedale, Wyoming, and the nuclear-disarmament advocates of the Global Security Institute; artist-activists such as Chester Arnold; and many other original thinkers. Theirs is a vision we might think of as *green patriotism*.

The term "green" is, admittedly, problematic; it springs directly from European aesthetic traditions that privileged verdant landscapes over all others and led to the desert-as-wasteland worldview (and, by contrast, the greenwashed desert in the opening scene of the Marine Corps recruitment commercial). Despite the word's baggage, I use "green" because it is an internationally recognized term for concern with environmental caretaking. Like popular patriotism, green patriotism is rooted in the landscape, but it is not beholden to warrior ideals or to long-term environmental exploitation for the sake of short-term financial gains. Green patriotism takes "love of country" literally—and by "country" I mean the messy breadth of the earth and not some pastoral fantasyland. Because green patriotism operates within a "moral ecology," it acknowledges that behaviors in one place affect people elsewhere. Therefore, it identifies "environmentalism," in the words of Robert F. Kennedy Jr., as a "civil rights issue" and environmental problems as social problems.[18] A patriotism of caretaking and community, green patriotism defies what Cheryll Glotfelty calls "placist" prejudices that classify some places as inferior to others, and it recognizes that even the word "ecology" implies living in a global community.

"We are creating, as did the Native Americans long before us, a patriotism based on ecological knowledge, moral consideration, ethical principle, spiritual belief, and a profound love for the earth underfoot," Nelson explains. "I believe this is the most basic, most urgent, and most vital patriotism of all, because conservationists are working in service to the elemental roots of their existence, as human organisms, as members of their communities, and as citizens of their nation's land."[19]

"Real patriots . . . know patriotism is about building decent and prosperous communities and protecting the soils, forests, water, and wildlife as the rightful legacy of our children and theirs," Orr tells us. "And they know the ancient truths that violence in all its forms is wrong and ultimately self-defeating, that health, holy, healing, and wholeness are one and indivisible."[20] I in no way mean to diminish the sacrifices, past and present, of American troops and their families. But there are facets to American stories of war and landscape, such as the national movements to conserve energy and to grow victory gardens during the twentieth century's World Wars, that get lost beside the fireworks of the military sublime and the mythology of war. Try to imagine what it would be like, in our times, to show love of country by growing a victory garden or by conserving energy. Try to imagine a patriotism that locates awe in acts of caretaking rather than in the shocking spectacles of the military sublime, the environmental brutality of the techno-industrial sublime, or the hydrogen- and fossil-fed dreams of the virtual sublime.

"I wish our national anthem were not the one about the bombs bursting in air, but the one about purple mountain majesties and amber waves of grain," writes Barbara Kingsolver in a remark that counters American military mythology with nineteenth-century landscape aesthetics.[21] What Kingsolver suggests is a shift in perspective that keeps the best of American aesthetic traditions—respect for land, reverence for place, gratitude for the luck or blessing that landed us in such a bountiful country and in a democracy—and reevaluates those mythologies that no longer serve us so well. That would be a sea change that would also identify, as Orr does, environmental problems as "the national-security issue of the early 21st century."[22] It could make our story the "caretaking" that Kittredge recommends.

On Earth as It Is in Heaven

"The American landscape was, to a great degree until the middle of the nineteenth century, an extraordinary opportunity to appreciate this world, and it evaporated so quickly," the California painter Chester Arnold observes.[23] Just as William Kittredge's essays challenge the frontier mythology that still informs American land use policies, much of Arnold's art engages in wry or raucous conversation with the works of his nineteenth-century predecessors. "My ambitions as an artist are mediated by my passions for the history and craft of painting as well as the political and environmental evolution of our species," he says. "I suppose these concerns place me in something of a romantic tradition—pursuing the sublime through the degradation [of our environment] and serving as witness and conscience to a culture moving too fast to notice its effects."[24]

Using his own version of nineteenth-century landscape aesthetics to talk back to Manifest Destiny aesthetics, Arnold depicts what so often gets left outside the frame—open pit mines, landfills, war zones, decaying urban neighborhoods, and postwar reconstruction. Dozens of Arnold's paintings address the afterlives of our consumption and the environmental effects of resource extraction. "Natural landscapes are subverted by [Arnold's] preoccupation with the detritus of human accumulation—cast off tires co-mingle with riverbeds and a rainbow springs up from the filth of a hog pit," observes one critic.[25]

Arnoldean sublimity is infused with accountability. Like Thomas Cole's *Oxbow* and *Course of Empire*, Arnold's canvases insist—albeit more directly than Cole's did—that viewers question themselves about the state of the world and their participation in environmental degradation. "These are stories, oblique narratives, of the human condition and how we colonize our world," Arnold says of his paintings, adding that his works are imbued with "a moralistic or even didactic subtext."[26]

Arnold's *On Earth As It Is In Heaven* (1997; fig. 6) is like a modern-day extension of Cole's argument in *The Oxbow* and a response to the Manifest Destiny canvases that Bierstadt, Moran, Emanuel Gottlieb

Leutze, and other artists created after Cole's death. Cole appreciated the beauty of the pastoral, but he was deeply disturbed by the excessive logging around his Catskills home. His *Oxbow*, with its question-mark river, its tangled forest ceding ground to a pastoral Promised Land, and its mountains inscribed with the Hebrew letters for "the Almighty," reflects his concern for the region's disappearing wilderness.[27] And if, as Goetzmann and Goetzmann insist, many nineteenth-century Americans saw Moran's *Mountain of the Holy Cross* as proof of God's blessing on the national project of Manifest Destiny, Chester Arnold's *On Earth As It Is In Heaven* shows us what that project has cost.

Arnold says that *On Earth As It Is In Heaven* was inspired by mine sites in Utah and Montana. At 3,960 feet deep, Kennecott Utah Copper's Bingham Canyon Mine is the world's oldest and largest open pit mine; the company boasts that it has produced "more copper than any mine in history—more than 17 million tons," and its equipment "removes 500,000 tons of material every day."[28] According to the Sierra Club, the mine has also produced "one of the world's most severe mining-generated ground water pollution" problems.[29] It's a mess that Utah's Department of Environmental Quality expects will take forty years to treat.[30] Butte, Montana, is home to ARCO's Berkeley Pit, a Superfund site and former copper mine that is "the nation's largest body of contaminated water."[31] At 700 feet deep and approximately one mile wide, the Berkeley Pit—a spectacular example of human engineering on the scale of the mountains that surround it—embodies the technological sublime. The water in the Berkeley Pit is so deadly that in 1995, a flock of 342 geese died when they paused there during migration.[32]

En route to establish the Massachusetts Bay Colony in 1630, John Winthrop had advised the faithful aboard the *Arbella* that their goal was to found a community based on Christian brotherhood, a "city on a hill" to serve as a beacon to the world. Nineteenth- and early twentieth-century developers called Butte, once the nation's largest producer of copper, "the richest hill on earth."[33] In Arnold's *On Earth As It Is In Heaven*, the city on the richest hill on earth is industrial, and the new cathedral is a factory. Toward the right of the canvas, atop one of the mine's mountains, an industrial plant in the shape of a Gothic cathedral emits puffs of smoke, the highest of which forms

the ghostly outline of a human head. Dense conifers cover the top
of the mountain and fringe both edges of the canvas. The sides of
the mine's benches look like tree bark, and the benches themselves
resemble stacked slices of sawn wood, suggesting that Arnold is ask-
ing us to think about the trees that were logged out as the mine was
carved into the earth.

Arnold's mine is a busy operation; dump trucks haul loads and
workers speckle the benches. In traditional sublime fashion, people
are minuscule, dwarfed by the landscape around them but also by the
structures they have built. Fires explode out of control. Hudson River
school paintings often featured a dead tree in the side foreground. But
instead of Cole's dead tree or Moran's geological cross, Arnold has
placed a conspicuous cruciform utility pole in the right foreground of
the painting. Similar utility poles line the benches, so that the entire
mine is staked with crosses, reminders that this is where Manifest
Destiny's religious and utilitarian motivations—the "meager utilitarianism" that Thomas Cole denounced—have taken us.

Beneath the most prominent cross, Arnold has painted the words
"The Axe Forgets / The Tree Remembers." A wire from that utility
pole draws the eye diagonally downward to a culvert that looks very
like a severed artery as it spews red fluid into the excavation. The solu-
tion in the holding pit is bright red, as if the mine were flooded with
blood—the blood, perhaps, of the earth, of the Native Americans
forced from the land, of the laborers killed on the job in the early days
of American industrialization, of all who suffer the effects of environmental contamination. Tires, detritus, and a small motorboat drift in
this toxic moat that encircles the cathedral mountain and curves out of
sight, suggesting, like the river in *The Oxbow*, a question mark. In the
left foreground, a dense cloud partially obscures lines of Arnold's
poetry that he has painted on the canvas, but the viewer can still make
out the words: "Forever bereft / one fate is left among / the leavings."
Our perspective on the scene is a God's-eye view; we look down on
the mine from the imperial position of a deity.

Arnold has titled several of his paintings with lines from the Lord's
Prayer because, he notes, "I've always found [the lines] so beautiful, and
they resonated with irony as I imagined and looked at certain landscapes.
And it seemed to me very poignant that some of the forces that were

leading to so much destruction of the landscape of the West in Manifest Destiny were oftentimes the same people who were so devout." (The title of this painting prompts the question: Whose will is being done?) Arnold hopes that his paintings will spur people to think about the effects of the resources they use and the goods they consume. "Painting is such a beautiful language," he says. "The whole purpose and meaning of painting is to create a dialogue for people who are moving so fast they don't have time to contemplate the effects of their actions. I hope to remind them it's worth taking the time."

"What would a paradise on earth be like?" William Kittredge muses. "Start with a process, I think, with everybody involved, taking part in the reimagining, thinking up the land of our heart's desiring, how things could be if cherishing were our main concern."[34] The keywords here are "everybody," "land," and "cherishing"; they speak of a community effort that eschews exploitation in favor of nurturing. John Tirman states that our next steps must be a collective effort. "Meeting the environmental challenge requires more than colossal investments in science and intensive diplomacy," he observes, adding, "it mandates a shift in the way we think about U.S. goals, our range of action, and our commitment to values beyond self-enrichment."[35]

Maurizio Viroli distinguishes between patriotism and nationalism by parsing their underlying values. "For the patriots," he explains, "the primary value is the republic and the free way of life that the republic permits; for the nationalists, the primary values are the spiritual and cultural unity of the people."[36] Green patriotism would not reduce diversity of opinion to partisan binaries of red and blue but would recognize and cherish rich and varied shades of green.

Emerson's Mirror

"A mythology," Kittredge notes, "can be understood as a story that contains a set of implicit instructions from a society to its members, telling them what is valuable and how to conduct themselves if they are to preserve the things they value."[37] I'm not oblivious to the irony—some would say hypocrisy, and perhaps rightfully so—of framing with a series of long drives a narrative that ultimately calls for conservation. My journey covered thousands of miles and far more places than the few I discussed here. But I believe it is important to investigate firsthand the places I write about, and at this point, I know of no better means of reaching them. At the Nevada Test Site, visitors were confined inside a tour bus except for one or two designated stops where we were permitted to disembark. Elsewhere, I spent as much time as I could outside my vehicle, exploring my surroundings.

Everywhere I looked, I smacked up against my own Emersonian mirror as I projected my perceptions, my past, and my hopes for the future onto the landscape. Unlike Emerson, I did not attempt to transcend the surface in search of the divine. Rather, I tried, as I have tried here, to shatter the lens of nineteenth-century landscape aesthetics by examining the rhetoric and iconography of landscape representations. As John Tirman observes, facing our environmental problems "requires a new lens on the world, one that sees in developing countries not bounty but common needs and aspirations."[38] I have sought to increase public awareness of the ways in which nineteenth-century mythology fosters "placism" in contemporary culture: by overlooking acts of environmental violence or by aestheticizing them as pastoral, picturesque, beautiful, or sublime, and by framing them in the rhetoric of the frontier and patriotic narratives of progress.[39]

"If living in history means that we cannot help leaving marks on a fallen world, then the dilemma we face is to decide what kinds of marks we wish to leave," William Cronon tells us.[40] The scale of environmental violence is staggering, and that's only what's obvious on the surface. Yet the scientists, artists, writers, community activists, and other green patriots I encountered in person and on paper convinced me that we will ultimately measure the value of our home in more than bombs, board feet, barrels of oil, and cubic feet of gas.

Scenic Cubicle

Tacked to the wall around my desk, I keep pictures of Montana that I cut from a calendar a friend gave to me when I moved out-of-state. There's a field of amber waves of grain leading up to a butte that looks much like one I hiked with two friends one hot September day.

There's an image of Ear Mountain, which neighbors the ranch where we worked, although it's shot from an angle that excludes the nearby nuclear missile silo. There's Grinnell Peak reflecting golden light at Glacier International Peace Park. And there's a picture of Yellowstone bison hoofing it through deep snow while plumes of steam from Old Faithful billow across the blue sky. This picture was likely taken in fat years before the great Yellowstone bison slaughter of the mid-1990s, when the snow fell so deep that the bison couldn't graze and they left the refuge of the park to search for forage. The Department of Agriculture doesn't permit bison to roam beyond the park for fear they'll spread brucellosis to Montana's cattle. But borders don't count for much among bison, and those that crossed to look for grass that winter were slaughtered or hazed back into the park, where they starved. Their story isn't over; the Department of Agriculture routinely harasses and slaughters bison that wander out of the park, but my picture of the Yellowstone buffalo doesn't show that part of the story.

I keep the pictures there because they remind me of a place I love. But I also keep them as reminders that representations of American nature are always far more complicated than a glance at the surface reveals. So often they depict the stories of an ecstatic nation in love with its landscape myths, reflected back in the Emersonian mirror. The other stories are etched into the earth, outside the limits of the frame.

NOTES

Introduction

1. Richard Slotkin, *The Fatal Environment: The Myth of the Frontier in the Age of Industrialization, 1800–1890* (New York: Harper Perennial, 1994), 47.
2. For an incisive discussion of the displacement of Native peoples from Yellowstone National Park, see Angela Miller, "The Fate of Wilderness in American Landscape Art: The Dilemmas of 'Nature's Nation,'" in *A Keener Perception: Ecocritical Studies in American Art History*, ed. Allan C. Braddock and Christoph Irmscher (Tuscaloosa: University of Alabama Press, 2009), 85–109. For a comprehensive review of recent controversies surrounding the slaughter of Yellowstone bison, see Matthew Brown, "Indian, Environmental Groups File Suit to Stop Yellowstone Bison Kills," *Missoulian*, November 9, 2009, available at http://missoulian.com.
3. "Louisiana Loses Population; Arizona Edges Nevada as Fastest Growing State," U.S. Census Bureau News Report, December 22, 2006, www.census.gov/newsroom/releases/archives/population/cb06-187.html; U.S. Census Bureau, *Population Change and Distribution 1990 to 2000* (Washington, D.C.: U.S. Department of Commerce, 2001).
4. Leo Marx, *The Machine in the Garden: Technology and the Pastoral Ideal in America* (1964; repr., New York: Oxford University Press, 2000), 195.
5. Robert F. Kennedy Jr., *Crimes against Nature: How George W. Bush and His Corporate Pals Are Plundering the Country and Hijacking Our Democracy* (New York: HarperCollins, 2004), 4.
6. Angela Miller, *The Empire of the Eye: Landscape Representation and American Cultural Politics, 1825–1875* (Ithaca, N.Y.: Cornell University Press, 1993); Barbara Novak, *Nature and Culture: American Landscape Painting, 1825–1875*, rev. ed. (New York: Oxford University Press, 1995).
7. Merle Curti, *The Roots of American Loyalty* (New York: Russell & Russell, 1946), vi–vii.

8. Cheryll Glotfelty, "Introduction: Literary Studies in an Age of Environmental Crisis," in *The Ecocriticism Reader*, ed. Glotfelty and Harold Fromm (Athens: University of Georgia Press, 1996), xviii–xix; Greg Garrard, *Ecocriticism* (London: Routledge, 2004), 4.

9. Braddock and Irmscher, *A Keener Perception*, 4, 9.

10. David Peterson del Mar, *Oregon's Promise: An Interpretive History* (Corvallis: Oregon State University Press, 2003), 9.

11. William Kittredge, *Who Owns the West?* (San Francisco: Mercury House, 1996), 70.

12. Scott Slovic, *Seeking Awareness in American Nature Writing: Henry Thoreau, Annie Dillard, Edward Abbey, Wendell Berry, and Barry Lopez* (Salt Lake City: University of Utah Press, 1992), 169.

1. Pass the Bottle: Scenic America

1. James Hill of the Archdiocese of Newark, New Jersey, provided this estimate through an intermediary; the archdiocese was unable to provide an exact count.

2. Jonathan Miller, "For 75 Years, It Was a Sight to Steer By in Newark," *New York Times*, June 27, 2006.

3. John O'Sullivan, "Annexation," in *Five Hundred Years: America in the World*, ed. Scott E. Casper and Richard O. Davies (Boston: Pearson, 2005), 142.

4. Amy DeRogatis, *Moral Geography: Maps, Missionaries, and the American Frontier* (New York: Columbia University Press, 2003), 2, 182.

5. Carolyn Merchant, "Reinventing Eden: Western Culture as Recovery Narrative," in *Uncommon Ground: Toward Reinventing Nature*, ed. William Cronon (New York: Norton, 1995), 133.

6. Barbara Novak, *Nature and Culture: American Landscape Painting, 1825–1875*, rev. ed. (New York: Oxford University Press, 1995), 16.

7. Ibid., 3.

8. Thomas Cole, "Essay on American Scenery," in *The American Landscape*, ed. John Conron (New York: Oxford University Press, 1974), 569.

9. Novak, *Nature and Culture*, 15.

10. Sue Rainey, *Creating Picturesque America: Monument to the Natural and Cultural Landscape* (Nashville: Vanderbilt University Press, 1994), xiii.

11. Angela Miller, *The Empire of the Eye: Landscape Representation and American Cultural Politics, 1825–1875* (Ithaca, N.Y.: Cornell University Press, 1993), 7–8.

12. Terry Gifford, *Pastoral* (London: Routledge, 1999), 15. Gifford's text provides a concise review of the various versions of the pastoral from ancient to contemporary times.

13. David Jacobson, *Place and Belonging in America* (Baltimore: Johns Hopkins University Press, 2002) 75–76.

14. Novak, *Nature and Culture*, 34; Judith K. Major, *To Live in the New World: A. J. Downing and American Landscape Gardening* (Cambridge: MIT Press, 1997), 48.

15. Anne Farrar Hyde, *An American Vision: Far Western Landscape and National Culture 1820–1920* (New York: New York University Press, 1990), 27.

16. Rochelle L. Johnson's *Passions for Nature: Nineteenth-Century America's Aesthetics of Alienation* (Athens: University of Georgia Press, 2009) provides a comprehensive review of aesthetic developments and their performance in nineteenth-century literature and culture in the United States.

17. Gunther Barth, "Timeless Journals: Reading Lewis and Clark with Nicholas Biddle's Help," *Pacific Historical Review* 16.4 (November 1994): 513, 515, 517. Barth notes that Biddle edited out the complexities of the experiences and many of the details that Lewis and Clark recorded, including their observations of the landscape and their interactions with Native peoples.

18. Blanche Linden-Ward, "Strange but Genteel Pleasure Grounds: Tourist and Leisure Uses of Nineteenth-Century Rural Cemeteries," in *Cemeteries and Gravemarkers: Voices of American Culture*, ed. Richard E. Meyer (Ann Arbor: UMI Research Press, 1989), 293.

19. Cole, "Essay on American Scenery," 571.

20. Ralph Waldo Emerson, *Selected Writings of Ralph Waldo Emerson*, ed. William H. Gilman (New York: Signet, 1965), 193.

21. Emerson, *Selected Writings*, 196.

22. Not all nineteenth-century landscape painters created works celebrating expansionism, and among those who did, such paintings were part—not all—of each artist's oeuvre. Many were more interested in interpreting nature's beauty than they were in endorsing or capitalizing on the construction of empire, and some of the most widely acclaimed paintings of Frederic Edwin Church and Martin Johnson Heade, for example, depicted landscapes or nature scenes outside of the United States. However, all of these works reflected and contributed to the overall popularity of landscape art.

23. Major, *To Live in the New World*, 91, 109.

24. Tamara Plakins Thornton, "Horticulture and American Character," in *Keeping Eden: A History of Gardening in America*, ed. Walter T. Punch (Boston: Little, Brown, 1992), 192–93.

25. Miller, *Empire of the Eye*, 9.

26. Ibid., 128.

27. Rainey, *Creating Picturesque America*, xiii.

28. Cole, "Essay on American Scenery," 570.

29. Emerson, *Selected Writings*, 194.

30. Scott Slovic, *Seeking Awareness in American Nature Writing: Henry Thoreau, Annie Dillard, Edward Abbey, Wendell Berry, and Barry Lopez* (Salt Lake City: University of Utah Press, 1992), 21.

31. Miller, *Empire of the Eye*, 11.

32. Ibid., 13. We might think of such images as what Louis Althusser called "Ideological State Apparatuses," his rather awkward term for cultural institutions, like politics and the arts, that worked to endorse the state's particular worldview. Louis Althusser, "Ideology and Ideological State Apparatuses," in *The Norton Anthology of Theory and Criticism*, ed. Vincent B. Leitch (New York: Norton, 2001), 1476–1508.

33. Miller, *Empire of the Eye*, 10.
34. William H. Goetzmann and William N. Goetzmann, *The West of the Imagination* (New York: Norton, 1986), 182.
35. Novak, *Nature and Culture*, 38.
36. John Muir, *My First Summer in the Sierra* (1911; repr., New York: Penguin, 1987), 101, 133, 208, 132.
37. Miller, "For 75 Years, It Was a Sight."
38. Leo Marx, *The Machine in the Garden: Technology and the Pastoral Ideal in America* (1964; repr., New York: Oxford University Press, 2000), 295.
39. Julia B. Corbett, "A Faint Green Sell: Advertising and the Natural World," in *Enviropop: Studies in Environmental Rhetoric and Popular Culture*, ed. Mark Meister and Phyllis M. Japp (Westport, Conn.: Praeger, 2002), 143.
40. Richard Slotkin, *Gunfighter Nation: The Myth of the Frontier in Twentieth-Century America* (Norman: University of Oklahoma Press, 1998), 6.
41. John Tirman, "The Future of the American Frontier," *American Scholar* 78.1 (Winter 2009): 30–40.
42. Robert F. Kennedy Jr., interview by Oprah Winfrey, "Interview: Oprah Talks to Bobby Kennedy, Jr.," *O*, February 2007, 174.
43. Yi-Fu Tuan, *Passing Strange and Wonderful: Aesthetics, Nature, and Culture* (Washington, D.C.: Island Press, 1993), 149.
44. William F. Zimmer, *Frontier Soldier: An Enlisted Man's Journal of the Sioux and Nez Perce Campaigns, 1877* (Helena: Montana Historical Society Press, 1998), 140, 153.
45. Arnold Berleant, *Aesthetics and Environment: Variations on a Theme* (Burlington, Vt.: Ashgate, 2005), 104.
46. Carson Walker, "Fund a Boon to Business" *Rapid City [N.D.] Journal*, April 2, 2005.
47. U.S. Department of the Interior, National Park Service, "Badlands Human History," May 28, 2005, www.nps.gov/badl/exp/humans.htm.
48. William Cronon, "The Trouble with Wilderness; or, Getting Back to the Wrong Nature," in *Uncommon Ground: Toward Reinventing Nature*, ed. Cronon (New York: Norton, 1995), 69.
49. Ibid., 72.
50. Ibid., 85.
51. Lawrence Buell, *The Environmental Imagination: Thoreau, Nature Writing, and the Formation of American Culture* (Cambridge: Belknap Press of Harvard University Press, 1995), 4.
52. Cheryll Glotfelty, "Literary Place Bashing, Test Site Nevada," in *Beyond Nature Writing: Expanding the Boundaries of Ecocriticism*, ed. Karla Armbruster and Kathleen Wallace (Charlottesville: University of Virginia Press, 2001), 243.

2. In the Name of the Bomb: The Wasteland's Atomic Bloom

1. M. G. Lord, *Forever Barbie: The Unauthorized Biography of a Real Doll* (New York: Avon, 1994), 8, 25, 38.

2. Terrence R. Fehner and F. G. Gosling, *Origins of the Nevada Test Site*, report prepared for U.S. Department of Energy, 2002, 82; A. Costandina Titus, *Bombs in the Backyard: Atomic Testing and American Politics* (Reno: University of Nevada Press, 2001), xv.

3. David J. Tietge, *Flash Effect: Science and the Rhetorical Origins of Cold War America* (Athens: Ohio University Press, 2002), 147.

4. Michael Scheibach, *Atomic Narratives and American Youth: Coming of Age with the Atom, 1945–1955* (Jefferson, N.C.: McFarland, 2003), 4.

5. Titus, *Bombs in the Backyard*, 20. It is uncertain for whom Little Boy was named.

6. Scheibach, *Atomic Narratives*, 4; James N. Yamazaki and Louis B. Fleming, *Children of the Atomic Bomb: An American Physician's Memoir of Nagasaki, Hiroshima, and the Marshall Islands* (Durham, N.C.: Duke University Press, 1995), 86.

7. U.S. Department of Energy, Nevada Operations Office, *United States Nuclear Tests, July 1945 through September 1992*, DOE/NV209-REV-15, December 2000, vii.

8. Mark Peterson, "Naming the Pacific," *Common-Place* 5.2 (January 2005): 3.

9. Titus, *Bombs in the Backyard*, 38.

10. Deborah G. Felder, *A Century of Women: The Most Influential Events in Twentieth Century Women's History* (Secaucus: Birch Lane Press, 1999), 187–88.

11. Titus, *Bombs in the Backyard*, 36.

12. Ibid., 37–54.

13. Fehner and Gosling, *Origins of the Nevada Test Site*, 40.

14. Cheryll Glotfelty, "Literary Place Bashing, Test Site Nevada," in *Beyond Nature Writing: Expanding the Boundaries of Ecocriticism*, ed. Karla Armbruster and Kathleen Wallace (Charlottesville: University of Virginia Press, 2001), 234.

15. Nevada's Yucca Mountain, which falls within the domain of the U.S. Department of Energy, was designated as the nation's future nuclear waste dump in 1987. In late 2009, the federal government failed to fund the project for much more than a license application to the Nuclear Regulatory Commission "to build a permanent dump" for nuclear waste. See Stephanie Kishi, "Yucca Mountain," *Las Vegas Sun*, November 27, 2010; Lisa Mascaro, "Small Space No Problem for Anti-Yucca Mountain Stand," *Las Vegas Sun*, October 16, 2009; "Life After Yucca Mountain," *Las Vegas Sun*, November 11, 2009.

16. Anne Farrar Hyde, *An American Vision: Far Western Landscape and National Culture, 1820–1920* (New York: New York University Press, 1990), 7.

17. Glotfelty, "Literary Place Bashing," 236.

18. Psalm 107:4–5 (New International Version).

19. Luke 4:2–13 (New International Version).

20. Ralph Waldo Emerson, *Selected Writings of Ralph Waldo Emerson*, ed. William H. Gilman (New York: Signet, 1965), 208.

21. State Bureau of Immigration, *Nevada and Her Resources* (Carson City, Nev.: State Printing Office, 1894).

22. Psalm 65:12–13 (New International Version).
23. Colleen Curran Griego et al., *The Nevada Test Site: A National Experimental Center* (Washington, D.C.: U.S. Government Printing Office, 1994), 2.
24. Matthew Coolidge, *The Nevada Test Site: A Guide to America's Nuclear Proving Ground* (Culver City, Nev.: Center for Land Use Interpretation, 1996), 30.
25. Titus, *Bombs in the Backyard*, 75–76.
26. Fehner and Gosling, *Origins of the Nevada Test Site*, 2.
27. Titus, *Bombs in the Backyard*, 97.
28. U.S. Department of Energy, "Miss Atom Bomb," *Nevada Test Site History*, www.nv.doe.gov/library/factsheets/DOENV_1024.pdf.
29. Ibid., 2.
30. Fehner and Gosling, *Origins of the Nevada Test Site*, 82.
31. Coolidge, *Nevada Test Site*, 27.
32. Ibid., 15.
33. Fehner and Gosling, *Origins of the Nevada Test Site*, 83.
34. Coolidge, *Nevada Test Site*, 26.
35. For a comprehensive review of the subsequent history of these soldiers, and that of the Nevada Test Site's downwinders, see Titus, *Bombs in the Backyard*.
36. U.S. Department of Energy, National Nuclear Security Administration, Nevada Operations Office, "Typical American Community Destroyed at Test Site," www.nv.doe.gov/library/publications/newsreviews/apple.htm.
37. Bob Hicok, "Bars Poetica," *Poets & Writers*, March/April 2004, 45.
38. U.S. Department of Energy, Nevada Operations Office, *United States Nuclear Tests, July 1945 through September 1992*, DOE/NV209-REV-15, December 2000, 28–32.
39. Coolidge, *Nevada Test Site*, 35.
40. "Summary of Vietnam Casualty Statistics," www.ktroop.com/HonorRoll/casualty.pdf; "Learn about the Vietnam War," *Digital History*, www.digitalhistory.uh.edu/modules/vietnam/index.cfm.
41. Rebecca Solnit, *Savage Dreams: A Journey into the Landscape Wars of the American West* (1994; Berkeley: University of California Press, 1999), 47.
42. Annette Kolodny, *The Lay of the Land: Metaphor as Experience and History in American Life and Letters* (Chapel Hill: University of North Carolina Press, 1975), 4.
43. Anthony Swofford, *Jarhead: A Marine's Chronicle of the Gulf War and Other Battles* (New York: Scribner, 2003), 114.
44. Emily Dickinson, *The Poems of Emily Dickinson*, Reading Edition, ed. R. W. Franklin (Cambridge: Harvard University Press, 1999), 269.
45. Steven Newcomb, "Bush at Mount Rushmore, 'The Shrine of Hypocrisy,'" *Indian Country Today*, August 6, 2004, available at www.indiancountry.com.
46. David W. Orr, *The Last Refuge: Patriotism, Politics, and the Environment in an Age of Terror* (Washington, D.C.: Island Press, 2004), 58.
47. William Kittredge, *Who Owns the West?* (San Francisco: Mercury House, 1996), 70.
48. Brian Bahouth, "Requiem for a Lake," *Reno News & Review*, April 29, 2004.
49. "Study: Genetics, Contaminants Could Be Cause of Fallon Cluster," Associated Press, *Las Vegas Sun*, October 1, 2005.

50. Korea Institute of Military History, *The Korean War*, 3 vols. (Lincoln: University of Nebraska Press, 1999), 3:768–69.

51. "Korean War Casualties," American Battle Monuments Commission, www.abmc.gov/wardead/listings/korean_war.php.

52. Walter Pincus, "Nuclear Strikes Part of Pre-Emptive Plans," *Reno Gazette Journal*, September 11, 2005.

53. Hans M. Kristensen, "Pentagon Cancels Controversial Nuclear Doctrine," Nuclear Information Project, February 2, 2006, www.nukestrat.com/us/jcs/canceled.htm; Launce Rake, "Bush's Denial of Plans for Iran Hit Wrong Chord Before Test Site Blast," *Las Vegas Sun*, April 11, 2006.

54. George Lipsitz, "Dilemmas of Beset Nationhood," in *Bonds of Affection: Americans Define Their Pariotism*, ed. John Bodnar (Princeton: Princeton University Press, 1996), 256.

55. "It's a Victory for Now," *Las Vegas Sun*, February 25, 2007; Launce Rake, "Concerns over NTS Tests 'Premature,'" *Las Vegas Sun*, October 11, 2006.

56. "'Divine Strake': The Process Behind Naming Bombs," *All Things Considered*, National Public Radio, March 31, 2006; "Infinite Justice, Out—Enduring Freedom, In," *BBC News*, September 25, 2001; Eleanor Clift, "Fathers and Sons," *Newsweek*, December 8, 2006.

57. Jonathan Granoff, "A Moment to Seize," Global Security Institute, November 2009, www.gsinstitute.org/gsi/pubs/moment_2009.pdf.

58. Swofford, *Jarhead*, 172.

59. George P. Shultz, William J. Perry, Henry A. Kissinger, and Sam Nunn, "A World Free of Nuclear Weapons," *Wall Street Journal*, January 4, 2007, available at www.gsinstitute.org/docs/01_04_07_WSJ.pdf.

60. "Remarks by President Barack Obama, Hradcany Square, Prague, Czech Republic," April 5, 2009. All remarks by President Obama cited in this chapter are available at www.whitehouse.gov/briefing-room/speeches-and-remarks.

61. "Statement by George Shultz, William Perry, Henry Kissinger and Sam Nunn Regarding the United Nations Security Council Meeting on Nuclear Nonproliferation and Nuclear Disarmament," September 24, 2009, available at www.whitehouse.gov/briefing-room/statements-and-releases.

62. "Remarks by the President at the United Nations Security Council Summit on Nuclear Non-Proliferation and Nuclear Disarmament," September 24, 2009.

63. "Remarks by the President on the Announcement of New START Treaty," March 26, 2010.

64. "Remarks by President Barack Obama, Hradcany Square."

65. Phyllis Andersen, "The City and the Garden," in *Keeping Eden: A History of Gardening in America*, ed. William T. Punch (Boston: Little, Brown, 1992), 160.

66. James Carroll, interview by Steve Inskeep, *Morning Edition*, National Public Radio, May 30, 2006.

67. James Carroll, *House of War: The Pentagon and the Disastrous Rise of American Power* (Boston: Houghton Mifflin, 2006), 512.

3. Timber Culture: Scenic Oregon and the Aesthetics of Clearcuts

1. Malcolm Andrews, *Landscape and Western Art* (Oxford: Oxford University Press, 1999), 116. For an in-depth study of the Claude glass, see Arnaud Maillet, *The Claude Glass: Use and Meaning of the Black Mirror in Western Art* (New York: Zone Books, 2004).

2. "The National Scenic Byways Program," America's Byways Online, www .byways.org/learn.

3. Sue Rainey, *Creating Picturesque America: Monument to the Natural and Cultural Landscape* (Nashville: Vanderbilt University Press, 1994), xiii.

4. Angela Miller, *The Empire of the Eye: Landscape Representation and American Cultural Politics, 1825–1875* (Ithaca, N.Y.: Cornell University Press, 1993), 128.

5. Marguerite S. Shaffer, *See America First: Tourism and National Identity, 1880–1940* (Washington, D.C.: Smithsonian Institution Press, 2001), 178.

6. Ibid., 3, 4.

7. Miller, *Empire of the Eye*, 10.

8. Shaffer, *See America First*, 168, 291.

9. U.S. Department of Commerce, *A Proposed Program for Scenic Roads and Parkways, Prepared for the President's Council on Recreation and Natural Beauty* (Washington, D.C.: U.S. Government Printing Office, 1966), 1.

10. Ibid., 5.

11. Ibid., 39, 38, 113.

12. U.S. Department of Transportation, Federal Highway Administration, "History of Scenic Road Programs," www.fhwa.dot.gov/infrastructure /scenichistory.cfm.

13. David Peterson del Mar, *Oregon's Promise: An Interpretive History* (Corvallis: Oregon State University Press, 2003), 192.

14. "Samuel H. Boardman State Scenic Corridor," Oregon Parks and Recreation Department: State Parks, www.oregonstateparks.org/park_77.php; S. H. Boardman, Arthur R. Kirkham, Arthur Crookham, Francis Lambert, and Thornton T. Munger, "Oregon State Park System: A Brief History," *Oregon Historical Quarterly* 55 (September 1954): 183.

15. For a study of the Andrews Experimental Forest, see Jon Luoma, *The Hidden Forest: The Biography of an Ecosystem* (New York: Henry Holt, 1999).

16. David B. Lindenmayer and Jerry F. Franklin, *Conserving Forest Biodiversity: A Comprehensive Multiscaled Approach* (Washington, D.C.: Island Press, 2002), 5.

17. Peter J. Bryant, "Temperate Forests and Deforestation," *Biodiversity and Conservation: A Hypertext Book* (University of California Irvine, 2002), www .dbc.uci.edu/~sustain/bio65/lec15/b65lec15.htm.

18. Del Mar, *Oregon's Promise*, 160.

19. James Cooper, *Knights of the Brush: The Hudson River School and the Moral Landscape* (New York: Hudson Hills, 2000), 45.

20. Miller, *Empire of the Eye*, 47.

21. Cooper, *Knights of the Brush*, 30.

22. Miller, *Empire of the Eye*, 33.

23. Thomas Cole, "Essay on American Scenery," in *The American Landscape: A Critical Anthology of Prose and Poetry*, ed. John Conron (New York: Oxford University Press, 1974), 577–78.

24. George B. Emerson, quoted in H. W. S. Cleveland, *Landscape Architecture, as Applied to the Wants of the West; with an Essay on Forest Planting on the Great Plains* (1873), repr., ed. Daniel J. Nadenicek and Lance M. Neckar (Amherst: University of Massachusetts Press and Library of American Landscape History, 2002), 95–96.

25. Ralph Waldo Emerson, *Selected Writings of Ralph Waldo Emerson*, ed. William H. Gilman (New York: Signet, 1965), 210.

26. Anne Farrar Hyde, *An American Vision: Far Western Landscape and National Culture, 1820–1920* (New York: New York University Press, 1990), 64–65.

27. William G. Robbins, "The Social Context of Forestry: The Pacific Northwest in the Twentieth Century," in *American Forests: Nature, Culture, and Politics*, ed. Char Miller (Lawrence: University Press of Kansas, 1997), 195.

28. Richard A. Rajala, *Clearcutting the Pacific Rain Forest: Production, Science, and Regulation* (Vancouver: University of British Columbia Press, 1998), xviii.

29. Walt Whitman, *Complete Poetry and Collected Prose* (New York: Library of America, 1982), 352.

30. Charles F. Wilkinson, *Crossing the Next Meridian: Land, Water, and the Future of the West* (Washington, D.C.: Island Press, 1992), 121.

31. Del Mar, *Oregon's Promise*, 100.

32. Karen Arabas and Joe Bowersox, "Introduction: Natural and Human History of Pacific Northwest Forests," in *Forest Futures: Science, Politics, and Policy for the Next Century*, ed. Arabas and Bowersox (Lanham, Md.: Rowman & Littlefield, 2004), xxxiii, xxxiv.

33. John A. Kitzhaber, "Principles and Politics of Sustainable Forestry in the Pacific Northwest: Charting a New Course," in Arabas and Bowersox, *Forest Futures*, 233.

34. Charles J. Hanley, "Forests in Flames: Scientists See Warming Ties," MSNBC. com, July 21, 2006.

35. Ralph Waldo Emerson, *Selected Writings*, 210.

36. James Howard, *U.S. Timber Production, Trade, Consumption, and Price Statistics, 1965–2005*, report for U.S. Department of Agriculture, Forest Service (Research Paper FPL-RP-637), September 2007, available at www.fpl.fs.fed. us/documnts/fplrp/fpl_rp637.pdf; Char Miller, *Gifford Pinchot and the Making of Modern Environmentalism* (Washington, D.C.: Island Press, 2001), 197.

37. "Book Review, *Elements of Forestry*," *Nature* 127 (February 14, 1931): 232; Thomas H. Mawson, "Review," *Town Planning Review* 5.4 (1915): 325–28.

38. "Dean F. Franklin Moon Records," and "Acting Dean Nelson C. Brown Records," New York State Archives, available at www.archives.nysed.gov; Frederick Franklin Moon and Nelson Courtlandt Brown, *Elements of Forestry* (New York: Wiley, 1915), 1.

39. Moon and Brown, *Elements of Forestry*, 3–4, vii–viii, 2.

40. Ibid., 2.

41. Ibid., 3.

42. Andrew C. Revkin, "Report Tallies Hidden Costs of Human Assault on Nature," *New York Times*, April 5, 2005.

43. George W. Bush, "President's Remarks on Healthy Forests," August 21, 2003. All remarks and proclamations by President Bush cited in this chapter are available at http://georgewbush-whitehouse.archives.gov/news/.

44. George W. Bush, "National Forest Products Week, 2004" Proclamation, October 21, 2004; George W. Bush, "National Forest Products Week Proclamation," October 19, 2001.

45. For further reading on Gifford Pinchot, see Char Miller, *Gifford Pinchot and the Making of Modern Environmentalism* (Washington, D.C.: Island Press, 2001).

46. Ibid., 4.

47. Ibid., 109–10.

48. Joanne Kosuda-Warner, *Landscape Wallcoverings* (New York: Cooper-Hewitt National Design Museum, 2001), 73; Kevin J. Avery and Franklin Kelly, *Hudson River School Visions: The Landscapes of Sanford Robinson Gifford* (New Haven: Yale University Press, 2003), 77.

49. Miller, *Gifford Pinchot*, 31.

50. Kirk Johnson, "From a Woodland Elegy, A Rhapsody in Green; Hunter Mountain Paintings Spurred Recovery," *New York Times*, June 7, 2001.

51. Miller, *Gifford Pinchot*, 108–9.

52. Avery and Kelly, *Hudson River School Visions*, 42.

53. Ibid., 43, 45.

54. Andrew Wilton and Tim Barringer, *American Sublime: Landscape Painting in the United States, 1820–1880* (Princeton: Princeton University Press, 2002), 120.

55. Miller, *Gifford Pinchot*, 108.

56. Johnson, "From a Woodland Elegy."

57. Hector St. John de Crèvecoeur, "What Is an American?," from *Letters from an American Farmer*, in Conron, *American Landscape*, 130.

58. Wilkinson, *Crossing the Next Meridian*, 127.

59. Ibid., 128.

60. Howard, *U.S. Timber Production*, 2.

61. Ibid., 3.

62. Ibid., 5, 4, 1.

63. "President Bush Signs American Dream Downpayment Act of 2003," remarks by president, December 16, 2003.

64. Howard, *U.S. Timber Production*, 5.

65. Renee Montagne, "Bigger Houses Pull More Electricity for Cooling," *Morning Edition*, National Public Radio, August 2, 2006.

66. Bush, "President's Remarks on Healthy Forests."

67. Lewis L. Gould, *Lady Bird Johnson: Our Environmental First Lady* (Lawrence: University Press of Kansas, 1999), 50.

68. James B. Craig and I. M. Moore, "White House Conference on Natural Beauty," *American Forests* 71.6 (June 1965): 13.

69. Reuel Little Tree Injection Company, "Kill Worthless Trees," *American Forests* 71.4 (April 1965): 49.

70. "Tongass Timber Sale," *American Forests* 71.9 (September 1965): 42.

71. Darius M. Adams, Richard W. Haynes, and Adam J. Daigneault, *Estimated Timber Harvest by U.S. Region and Ownership, 1950–2002*, U.S. Department of Agriculture, Forest Service, Pacific Northwest Research Station, General Technical Report PNW-GTR-659, January 2006, 63; this report is available at www.fs.fed.us/pnw/pubs/pnw_gtr659.pdf. The authors report this quantity as 1,233,000,000 cubic feet. In my study, board feet were calculated based on a conversion factor of 183 cubic feet=1,000 board feet. Depending on the type of wood and the average diameter of the tree, the conversion factor can vary. This formula was used for all references cited from this report.

72. William B. Morse, "That Utilizing Monster," *American Forests* 71.1 (January 1965): 30.

73. Ibid.

74. William Kittredge, *Owning It All* (St. Paul: Graywolf Press, 1987), 62.

75. Virginia Scott Jenkins, *The Lawn: A History of an American Obsession* (Washington, D.C.: Smithsonian Institution, 1994), 135, 159.

76. Beatriz Colomina, "The Lawn at War: 1941–1961," in *The American Lawn*, ed. Georges Teyssot (New York: Princeton Architectural Press, 1999), 140.

77. U.S. Department of Energy, Nevada Operations Office, *United States Nuclear Tests, July 1945 through September 1992*, DOE/NV-209 (Rev. 14), December 1994, 23–26; a copy of this report is available at www.fas.org/nuke/guide/usa/nuclear/usnuctests.htm.

78. Mark E. Harmon, "Moving towards a New Paradigm for Woody Detritus Management," in *USDA Forest Service Gen. Tech. Rep. PSW-GTR-181* (2002), 932, 933, 930; available at www.fs.fed.us/psw/publications/documents/gtr181/071_Harm.pdf.

79. Ibid., 931, 932.

80. Rick Bass, *The Book of Yaak* (Boston: Mariner Books, 1996), 11; Harmon, "Moving towards a New Paradigm," 929.

81. Harmon, "Moving towards a New Paradigm," 933.

82. Frederick J. Swanson, "Long Term Ecological Reflections Plots" (The Andrews Experimental Forest LTER, 2002), http://andrewsforest.oregonstate.edu/research/related/writers/template.cfm?next=plots&topnav=169#logdecomp.

83. Bass, *Book of Yaak*, 11.

84. Harmon, "Moving towards a New Paradigm," 935.

85. Mary Ann Albright, "Damage Control after Logging Study?," *Corvallis Gazette Times*, February 5, 2006, available at www.gazettetimes.com.

86. Ibid.

87. U.S. Office of the President, *Healthy Forests: An Initiative for Wildfire Prevention and Stronger Communities* (White House Report, August 22, 2002), 6; this report is available at www.ruraltech.org/projects/fire/forest_fires/Healthy_Forests_v2.pdf.

88. George Wuerthner, "Bugged: Is the War on Beetles More Hype Than Substance?," *Forest Magazine*, Spring 2007, 39–40.

89. Ibid., 40.

90. "Ravenous Beetles Devastate Forest," *(Eugene, Ore.) Register-Guard*, October 31, 2005, www.registerguard.com.

91. Charles Petit, "In the Rockies, Pines Die and Bears Feel It," *New York Times*, January 30, 2007; Wuerthner, "Bugged," 41.

92. Wuerthner, "Bugged," 42.

93. Harmon, "Moving towards a New Paradigm," 942.

94. Robert Davis, "Into the Forest," in *Dancing on the Rim of the World: An Anthology of Contemporary Northwest Native American Writing*, ed. Andrea Lerner (Tucson: Sun Tracks and University of Arizona Press, 1990), 149.

95. Del Mar, *Oregon's Promise*, 221.

96. Ibid., 262.

97. Patrick Mazza, "The Mud Next Time: Northwest Clearcutting Is Producing Deadly Landslides," *Sierra*, May–June 1997, www.sierraclub.org /sierra/199705/priorities.asp.

98. Frederick Swanson quoted in Mazza, "The Mud Next Time."

99. Paul Shepard, *Man in the Landscape: A Historic View of the Esthetics of Nature* (1967; repr., Athens: University of Georgia Press, 2002), 87; Jenkins, *The Lawn*, 22, 187; Mark Wigley, "The Electric Lawn," in *The American Lawn*, ed. Georges Teyssot (New York: Princeton Architectural, 1999), 156; Colomina, "The Lawn at War," 149.

100. Emerson, *Selected Writings*, 210.

101. Peter Browning, *Yosemite Place Names* (Lafayette, Calif.: Great West Books, 1988), 21.

102. William H. Goetzmann and William N. Goetzmann, *The West of the Imagination* (New York: Norton, 1986), 182.

103. William Cullen Bryant, *Picturesque America; or, The Land We Live In: A Delineation by Pen and Pencil of the Mountains, Rivers, Lakes, Forests, Water-falls, Shores, Cañons, Valleys, Cities, and Other Picturesque Features of Our Country* (New York: D. Appleton, 1872), iii.

104. Miranda J. Green, *The World of the Druids* (London: Thames and Hudson, 1997), 109.

105. Luke 19:1–4 (New International Version).

106. William Cullen Bryant, "A Forest Hymn," in Conron, *American Landscape*, 292.

107. Emerson, *Selected Writings*, 193.

108. Henry David Thoreau, *Walden. Civil Disobedience*, ed. Sherman Paul (Boston: Houghton Mifflin, 1960), 139.

109. Ibid., 62.

110. Simon Schama, *Landscape and Memory* (New York: Knopf, 1995), 187, 190, 191.

111. Bryant, *Picturesque America*, iii.

112. John Muir, *My First Summer in the Sierra* (1911; repr., New York: Penguin, 1987), 41.

113. Ibid., 147.

114. Henry Lee Morgenstern, "Cutting the Cathedral: The War over the Future of Alaska's Tongass, America's Largest Old-Growth Forest," *E* 8.4 (July 1, 1997), 36.

115. Barry Lopez, *Crossing Open Ground* (1978; repr., New York: Scribner, 1988), 53.

116. Clark S. Binkley, "Forestry in the Long Sweep of History," in *Forest Policy for Private Forestry: Global and Regional Challenges*, ed. Lawrence Teeter (New York: CABI Publishing, 2003), 5–6.

117. "Andrews LTER," *Twenty Years of Research*, Long Term Ecological Research Network, www.lternet.edu/vignettes/and.html.

118. Thoreau quoted in David R. Foster, *Thoreau's Country: Journey through a Transformed Landscape* (Cambridge: Harvard University Press, 1999), 38.

119. Swanson, "Long Term Ecological Reflections Plots."

120. Ibid.

121. Frederick Swanson, "Roles of Scientists in Forestry Policy and Management: Views from the Pacific Northwest," in Arabas and Bowersox, *Forest Futures: Science, Politics, and Policy for the Next Century*, 115.

122. Swanson, "Long Term Ecological Reflections Plots."

123. Bass, *Book of Yaak*, 6.

124. Swanson, "Roles of Scientists in Forestry Policy and Management," 120.

125. Ibid., 122.

126. Constance Best and Laurie A. Wayburn, *America's Private Forests: Status and Stewardship* (Washington, D.C.: Island Press, 2001), 3.

127. Frank Luntz, *Words That Work: It's Not What You Say, It's What People Hear* (New York: Hyperion, 2007), 209–11.

128. U.S. Office of the President, *Healthy Forests*, 1.

129. Ibid., 3, 2.

130. Jaina L. Moan and Zachary A. Smith, "Bush and the Environment," in *A Bird in the Bush: Failed Policies of the George W. Bush Administration*, ed. Dowling Campbell (New York: Algora Publishing, 2005), 97–98.

131. Swanson, "Roles of Scientists in Forestry Policy and Management," 121.

132. Swanson, telephone interview by author, December 15, 2005.

133. Kenneth E. Foote, *Shadowed Ground: America's Landscapes of Violence and Tragedy* (Austin: University of Texas Press, 1997), 292.

134. Robert Leo Heilman, *Overstory: Zero: Real Life in Timber Country* (Seattle: Sasquatch, 1995), 22.

135. Wilkinson, *Crossing the Next Meridian*, 142.

136. Miller, *Gifford Pinchot*, 359.

137. Best and Wayburn, *America's Private Forests*, xxi.

138. Heilman, *Overstory: Zero*, 28–29.

139. Ibid., 30.

140. Bass, *Book of Yaak*, 37.

141. Lindenmayer and Franklin, *Conserving Forest Biodiversity*, 9.

142. Henry David Thoreau, *Walking*, 1862, in *The Portable Thoreau*, ed. Carl Bode (1947; repr., New York: Penguin, 1975), 609.

143. Lindenmayer and Franklin, *Conserving Forest Biodiversity*, 5.

144. Ibid., 15.

145. Kathleen Dean Moore, *The Pine Island Paradox: Making Connections in a Disconnected World* (Minneapolis: Milkweed Editions, 2004), 212–13.

146. Mark Schleifstein, "Northwest Salmon Fading Fast," part 6, *New Orleans Times-Picayune*, March 29, 1996, available at www.pulitzer.org/archives/6033.

147. Schleifstein, "Northwest Salmon Fading Fast"; U.S. Department of the

Interior, National Park Service, "Salmon," *Salmon Conservation and Restoration Efforts in the Columbia Cascades*, February 18, 2004, www.nps.gov/ccso/salmonid.htm.

148. Heilman, *Overstory: Zero*, 76.

149. Adams, Haynes, and Daigneault, *Estimated Timber Harvest*, 63; converted from cubic feet with a conversion factor of 183.

150. Ibid., 28.

151. Howard, *U.S. Timber Production*, 4.

152. Ibid., 4.

153. Best and Wayburn, *America's Private Forests*, 16.

154. Ibid., xix–xx; Sherri Richardson Dodge, "Society's Choices: Land Use Changes, Forest Fragmentation, and Conservation," *Science Findings* 88 (November 2006): 1, available at http://www.fs.fed.us/pnw/sciencef/scifi88.dpf.

155. Best and Wayburn, *America's Private Forests*, xx.

156. Doge, "Society's Choices," 1.

157. Alig quoted ibid., 3.

158. Lindenmayer and Franklin, *Conserving Forest Biodiversity*, 15, 5.

159. Miller, *Empire of the Eye*, 11.

160. Ibid., 10.

161. Amy DeRogatis, *Moral Geography: Maps, Missionaries, and the American Frontier* (New York: Columbia University Press, 2003), 2, 95, 182.

162. Miller, *Empire of the Eye*, 10.

163. Malcolm Andrews, *Landscape and Western Art* (Oxford: Oxford University Press, 1999), 116.

164. "The National Scenic Byways Program," America's Byways Online, www.byways.org/learn/.

165. Maillet, *The Claude Glass*, 192.

4. Open (for Business) Range: Wyoming's Pay Dirt and the Virtual Sublime

1. W. Dale Nelson, "Dem Pushes Coal Study," *Casper (Wyo.) Star-Tribune*, July 12, 2006. All *Casper Star-Tribune* articles cited in this chapter were accessed at http://trib.com.

2. Dustin Bleizeffer, "Wyoming Coal Sales Decline," *Billings (Mont.) Gazette*, January 10, 2010. All *Billings Gazette* articles cited in this chapter were accessed at www.billingsgazette.com. Bob Moen, "Jonah, Pinedale Gas Fields Yield Oil, Too," *Casper Star-Tribune*, July 14, 2009; Mead Gruver, "Wyoming Gas, Oil Drilling Permits Decline," Associated Press, January 15, 2010, available at http://abcnews.go.com/Business/wireStory?id=9570299.

3. U.S. Department of Energy, Energy Information Administration, "Independent Statistics and Analysis: Top 100 U.S. Oil and Gas Fields By 2009 Proved Reserves," December 7, 2010, www.eia.gov/oil_gas/rpd/topfields.pdf; Phil Taylor, "Mule Deer Declines in Wyo. Gas Field Warrant 'Serious' Mitigation Response," *Casper Star-Tribune*, October 22, 2010.

4. Angela Miller, *The Empire of the Eye: Landscape Representation and American Cultural Politics, 1825–1875* (Ithaca, N.Y.: Cornell University Press, 1993), 7–8, 128.

5. Robert M. Utley, *A Life Wild and Perilous: Mountain Men and the Paths to the Pacific* (New York: Henry Holt, 1997), 120, 175, 192.

6. Nancy K. Anderson, *Thomas Moran* (New Haven: Yale University Press, 1997), 53, 54.

7. Ibid., 50.

8. Merle Curti, *The Growth of American Thought*, 3rd ed. (1964; repr., New Brunswick: Transaction, 1991), 395.

9. Jonathan M. Hansen, *The Lost Promise of Patriotism: Debating American Identity, 1890–1920* (Chicago: University of Chicago Press, 2003), 10, quoting James T. Kloppenberg.

10. John Bodnar, "The Attractions of Patriotism," in *Bonds of Affection: Americans Define Their Pariotism*, ed. Bodnar (Princeton: Princeton University Press, 1996), 11, 15, 16.

11. E. L. Burlingame, "The Plains and the Sierras," in *Picturesque America; or, The Land We Live In: A Delineation by Pen and Pencil of the Mountains, Rivers, Lakes, Forests, Water-falls, Shores, Cañons, Valleys, Cities, and Other Picturesque Features of Our Country*, ed. William Cullen Bryant (New York: D. Appleton, 1872), 174.

12. Ruth Lauritzen, "Nature's Art Shop," *Green River: Nature's Art Shop* (Green River, Wyo.: Green River Historic Preservation Commission, 2004), 1.

13. Mead Gruver, "Drilling Crossroads," *Casper Star-Tribune*, February 23, 2010.

14. Chris Merrill, "Too Many Wells on the Anticline?," *Casper Star-Tribune*, April 21, 2008; Ted Williams, "For a Week's Worth of Gas," *Mother Jones*, September/October 2004, 69; Jeff Gearino, "Operators Seek Year-Round Drilling," *Casper Star-Tribune*, June 23, 2005; Whitney Royster, "Industry Walks a Fuzzy Line between Preservation and Extortion," *High Country News*, August 8, 2005, www.hcn.org/issues/303/15685.

15. Carey Bylin et al., "Methane's Role in Promoting Sustainable Development in the Oil and Natural Gas Industry," October 5, 2009, www.epa.gov/gasstar/documents/best_paper_award.pdf.

16. Jeff Gearino, "Jonah Field: Biggest Gas-Producing Field Is Also Biggest Oil Producer," *Casper Star-Tribune*, May 29, 2008; Bob Moen, "Jonah, Pinedale Gas Fields Yield Oil Too," *Casper Star-Tribune*, July 14, 2009.

17. Dean P. DuBois et al., "Geology of Jonah Field" (abstract), EnCana Oil and Gas (USA), www.cspg.org/conventions/abstracts/2003abstracts/245S0130.pdf.

18. Jeff Gearino, "Power Plant, Coal Mine, Celebrate 30 Years," *Casper Star-Tribune*, August 21 2004.

19. Dustin Bleizeffer, "Panel Pushes CO_2 Business," *Casper Star-Tribune*, July 26, 2006.

20. Mae Urbanek, *Wyoming Place Names* (Missoula, Mont.: Mountain Press, 1988), 60.

21. Carolyn Merchant, "Reinventing Eden: Western Culture as Recovery Narrative," in *Uncommon Ground: Toward Reinventing Nature*, ed. William Cronon (New York: Norton, 1995), 133.

22. William Kittredge, *Who Owns the West?* (San Francisco: Mercury House, 1996), 74.

23. Virginia Scharff, "Women of the West," *History Now: American History Online* 9 (September 2006), www.gilderlehrman.org/historynow/09_2006/historian5. php; Mike Mackey, Introduction to *The Equality State*, ed. Mackey (Powell, Wyo.: Western History Publications, 1999), 1.

24. David E. Nye, *American Technological Sublime* (Cambridge: MIT Press, 1994), 147.

25. Leo Marx, *The Machine in the Garden: Technology and the Pastoral Ideal in America* (1964; repr., New York: Oxford University Press, 2000), 295.

26. Richard Slotkin, *Gunfighter Nation: The Myth of the Frontier in Twentieth-Century America* (Norman: University of Oklahoma Press, 1998), 63.

27. David E. Nye, quoted in Astrid Boger, "Envisioning Progress at Chicago's White City," in *Space in America: Theory, History, Culture*, ed. Klaus Benesch and Kerstin Schmidt (Amsterdam: Rodopi, 1999), 273.

28. Nye, *American Technological Sublime*, 147.

29. Wyoming Secretary of State, "Wyoming's Bucking Horse and Rider," http:// soswy.state.wy.us/AdminServices/BHROverview.aspx.

30. "The Day of the Cowboy," *American Cowboy*, July/August 2006, 32–33.

31. Benjamin F. Shearer and Barbara S. Shearer, *State Names, Seals, Flags, and Symbols* (Westport, Conn.: Greenwood, 1994), 67.

32. Whitney Royster, "Pronghorn Affected More by Habitat Fragmentation, Study Says," *Casper Star-Tribune*, June 18, 2006.

33. Jeff Gearino, "Sportsmen's Group Proposes Wildlife Coalition," *Casper Star-Tribune*, July 8, 2006.

34. Hall Sawyer and Ryan Nielson, *Mule Deer Monitoring in the Pinedale Anticline Project Area: 2010 Annual Report*, September 14, 2010, www.wy.blm.gov/jio -papo/papo/reports/2010annualreport_muledeer.pdf. The biologist Hall Sawyer, one of the report's coauthors, observes elsewhere that some of the deer may have declined to migrate during a series of warmer winters, or that some may have overwintered on other favorable terrain. See Taylor, "Mule Deer Declines in Wyo. Gas Field." See also Cat Urbigkit, "Pinedale Mesa Deer Population Drops," *Casper Star-Tribune*, October 28, 2010.

35. Geoffrey O'Gara, "Hostile Beauty," *National Wildlife Magazine*, August/September 2002, www.nwf.org.

36. Ted Williams, "For a Week's Worth of Gas," *Mother Jones*, September/October 2004, 70. The state responded to these stresses on habitat by establishing the Jonah Interagency Mitigation and Reclamation Office and the Pinedale Anticline Project Office; both offices represent cooperative efforts of the state's BLM and Departments of Agriculture, Environmental Quality, and Fish and Game. The offices are charged with "provid[ing] overall management of on-site monitoring and off-site mitigation activities primarily focusing on" each area's wildlife. See "Jonah Interagency Office–Pinedale Anticline Project Office," www.wy.blm.gov/jio-papo/index.htm.

37. O'Gara, "Hostile Beauty."

38. Ibid.; Jeff Gearino, "Companies Propose More Wamsutter Drilling," *Casper Star-Tribune*, March 7, 2006.

39. Ben Merchant Vorpahl, *My Dear Wister: The Frederic Remington–Owen Wister Letters* (Palo Alto, Calif.: American West, 1972), 18; Joan Carpenter Troccoli, *Painters of the American West: The Anschutz Collection* (New Haven: Yale University Press and Denver Art Museum, 2000), 137.

40. Chase Squires, "Rocky Mountain Sigh," *Casper Star-Tribune*, July 22, 2006.

41. Cory Hatch, "Feds Want More Wells Even as Deer Decline," *Jackson Hole News and Guide*, December 20, 2006, www.jhnewsandguide.com; Alecia Warren, "Air Issue Threatens County Tourism," *Pinedale Roundup*, April 17, 2008, www.sublette.com/roundup.

42. Whitney Royster, "EPA: Weather Didn't Cause Ozone Spikes," *Casper Star-Tribune*, February 8, 2006; Jeff Gearino, "EPA Tightens Ozone Rules," *Casper Star-Tribune*, January 8, 2010.

43. Jeff Gearino, "DEQ Issues Ozone Alert for Pinedale," *Casper Star-Tribune*, February 3, 2009.

44. Steve Hargreaves, "Small Town, Big Changes," CNN Money.com, October 17, 2008, money.cnn.com/2008/09/18/news/economy/wyoming_drilling/index.htm.

45. Whitney Royster, "BLM: Drilling Will Bring Haze," *Casper Star-Tribune*, August 11, 2005.

46. Chris Merrill, "Your Voice Is Not Really Heard," *Casper Star-Tribune*, September 22, 2008. Randy Teeuwen, a spokesman for the energy producer EnCana, said in early 2010 that, following the ozone alerts, Sublette County energy companies had modified equipment and reduced emissions by 60 percent. In early 2010, the Environmental Protection Agency proposed tightening standards for ozone limits nationwide. See "Industry Should Be Able to Meet New Ozone Rules," *Casper Star-Tribune*, January 11, 2010.

47. Dustin Bleizeffer, "Eye on Energy: New Energy Mix Requires Water," *Casper Star-Tribune*, December 20, 2009.

48. "'Fracking' Debate Heats Up," *Casper Star-Tribune*, June 11, 2009. The regulations would require physical buffers around new wells and installation of "backflow prevention devices" to prevent contaminated fluids from leaching into groundwater. The state engineer also recommended collection of baseline data on each newly drilled well's water quality. See "State Engineer to Enact New Water Well Rules in Wyoming," *Billings (Mont.) Gazette*, January 27, 2010. Concerns about hydraulic fracturing extend beyond Wyoming. Energy producers practice hydraulic fracturing in other states, and in February 2010 the House Energy and Commerce Committee began investigating the process in response to plans to drill the gas-rich Marcellus Shale, which underlies parts of New York, Pennsylvania, Ohio, and West Virginia. See Matthew Daly, "House Panel to Investigate Hydraulic Drilling," *Casper Star-Tribune*, February 21, 2010. For additional information on concerns about gas drilling, see Abrahm Lustgarten, "Natural Gas Drilling: What We Don't Know," *ProPublica: Journalism in the Public Interest*, December 31, 2009, www.propublica.org.

49. U.S. Department of Energy, Energy Information Administration, "Independent Statistics and Analysis: Wyoming," February 4, 2010, www.eia.gov/cfapps/state/state_energy_profiles.cfm?sid=WY.

50. Dustin Bleizeffer, "Keeping the Dust Down," *Casper Star-Tribune*, June 24, 2006.

51. Dustin Bleizeffer, "Report: Grouse Protections Not Working," *Casper Star-Tribune*, June 28, 2006. In March 2010, the Department of the Interior declined to list the greater sage grouse as either threatened or endangered under the Endangered Species Act. Ken Salazar, secretary of the interior, noted that listing the bird was "'warranted but precluded' . . . for now because other species are a higher priority." See Jim Tankersley, "Interior Hesitates on Sage Grouse," *Baltimore Sun*, March 5, 2010.

52. Clair Johnson, "Montana Defends Water Standards," *Casper Star-Tribune*, June 20, 2006.

53. "Wyoming Asks EPA to Discard Montana Rules," *Billings (Mont.) Gazette*, April 6, 2006; Matthew Brown, "Judge Overturns Montana Water Rules for Gas Drilling," *Missoulian*, October 14, 2009, available at http://missoulian.com.

54. "Wyoming Issues," *Casper Star-Tribune*, August 22, 2006.

55. Quoted in Whitney Royster, "Tip of a Melting Iceberg?," *Casper Star-Tribune*, July 17, 2006.

56. Ibid.

57. U.S. Department of Energy, "U.S. and South Korea Sign Agreement on FutureGen Project," June 26, 2006, www.energy.gov/news/archives/3778.htm; Rebecca Solnit, *Savage Dreams: A Journey into the Landscape Wars of the American West* (1994; Berkeley: University of California Press, 1999), 47.

58. Vincent Mosco, *The Digital Sublime: Myth, Power, and Cyberspace* (Cambridge: MIT Press, 2004), 4, 24.

59. Lee Rozelle, *Ecosublime: Environmental Awe and Terror from New World to Oddworld* (Tuscaloosa: University of Alabama Press, 2006), 1–2.

60. The U.S. Department of Energy has had an "on-off-on-again" relationship with the FutureGen Alliance; in June 2009 Secretary of Energy Steven Chu announced that the agency and the alliance were back "on." See Ben Geman and Greenwire, "DOE Revives FutureGen, Reversing Bush-Era Decision," *New York Times*, June 12, 2009.

61. U.S. Department of Energy, "FutureGen: Tomorrow's Pollution-Free Power Plant," *Research News*, www.eurekalert.org/features/doe/2006-09/detl -f-t_1092206.php.

62. Thomas Cole, "Essay on American Scenery," reprinted in *The American Landscape*, ed. John Conron (New York: Oxford University Press, 1974), 572.

63. FutureGen Industrial Alliance, "Frequently Asked Questions," February 2010, www.futuregenalliance.org/faqs.stm.

64. U.S. Department of Energy, "Clean Coal Technology and the President's Clean Coal Power Initiative," *Fossil Energy: DOE's Clean Coal Technology Program*, June 29, 2006, www.fossil.energy.gov (no longer available).

65. Paul Shepard, *Man in the Landscape: A Historic View of the Esthetics of Nature* (1967; repr., Athens: University of Georgia Press, 2002), 87.

66. Virginia Scott Jenkins, *The Lawn: A History of an American Obsession* (Washington, D.C.: Smithsonian Institution, 1994), 22, 185.

67. Beatriz Colomina, "The Lawn at War: 1941–1961," in *The American Lawn,* ed. Georges Teyssot (New York: Princeton Architectural Press, 1999), 149.

68. Mark Wigley, "The Electric Lawn," ibid., 156.

69. Noelle Straub, "Wyo Drops Out of Running for Plant," *Casper Star-Tribune,* July 26, 2006; "Wyo's Favored Site Wins Plant," *Casper Star-Tribune,* December 19, 2007.

70. Jeff Gearino, "Wyoming among Carbon Sequestration Pioneers," *Casper Star-Tribune,* November 27, 2009; Jeff Gearino, "Sequestration Plant Study Goes Public," *Casper Star-Tribune,* January 26, 2010; Jeff Gearino, "BLM Releases Study on Innovative Gas Plant," *Casper Star-Tribune,* January 25, 2010; Jeff Gearino, "UW, GE Teamed Up to Advance Coal Gasification," *Casper Star-Tribune,* November 27, 2009.

71. U.S. Department of Energy, "Coal," www.energy.gov/energysources/coal.htm.

72. George W. Bush, "President Discusses Advanced Energy Initiative in Milwaukee," February 20, 2006, available at http://georgewbush-whitehouse.archives.gov/news/releases.

73. Dustin Bleizeffer, "Wyo Offers Up $31.2 M Incentive for FutureGen," *Casper Star-Tribune,* June 11, 2006.

74. Just how clean is American coal? Outside the scope of this book but germane to any discussion of "clean coal" are the considerable social costs of mining and processing coal. The U.S. Department of Labor reports that 104,674 coal mining workers died in workplace accidents between 1900 and 2009; 318 of these workers died between 2000 and 2009, and an additional 35 miners were killed on the job during the first quarter of 2010. (In 1973, the Department of Labor began including deaths of office staffers in these tallies. These numbers do not reflect the deaths that occurred from occupational illnesses, such as black lung disease.) See U.S. Department of Labor, Mine Safety and Health Administration, "Coal Daily Fatality Report—April 30, 2010," www.msha.gov/stats/charts/coaldaily.asp; U.S. Department of Labor, Mine Safety and Health Administration, "Coal Fatalities for 1900–2009," www.msha.gov/stats/centurystats/coalstats.asp.

West Virginia, which supplied nearly 158 million tons of coal in 2008, is second only to Wyoming in coal production, and the Appalachian region of the United States, which produced more than 390 million tons of coal in 2008, is the nation's second most productive region. Mountaintop removal is a particularly aggressive form of surface coal mining practiced mainly in West Virginia, Kentucky, Tennessee, and Virginia; mountaintops are blasted off with dynamite and the rock is dumped into "valley fills," often poisoning headwater streams. In 2008, there were 1,278 active coal mines in Appalachia; of these, 745 were surface mines. I am unable to obtain an accurate count of mountaintop removal sites, but in 2006, John G. Mitchell reported in *National Geographic* that the process "has impacted more than 400,000 acres in this four-state Appalachian region." The U.S. Environmental Protection Agency "estimate[s] that almost 2,000 miles of Appalachian headwater streams have been buried by mountaintop coal mining." In April 2010, EPA administrator Lisa Jackson announced guidelines that would make it harder for companies to obtain permits to dump rock waste into valley fills. See U.S. Department of Energy, Energy Information Administration, *Annual Coal Report 2008*, September 18, 2009, www.eia.doe.gov/cneaf/coal/page/acr/tables2.pdf; U.S. Department of Energy, Energy Information Administration, "Coal Production and Number of Mines by State and Mine Type," September 18, 2009, www.eia.doe.gov/cneaf/coal/page/acr/table1.html; Appalachian Voices, "Mountaintop Removal Coal Mining," 2007, www.appvoices.org/index.php?/site/mtr_overview; John G. Mitchell, "When Mountains Move," *National Geographic*, March 2006, http://ngm.nationalgeographic.com; U.S. Environmental Protection Agency, "EPA Issues Comprehensive Guidance to Protect Appalachian Communities from Harmful Environmental Effects of Mountaintop Mining," April 1, 2010, www.epa.gov/aging/press/epanews/2010/2010_0401_1.htm.

Processing coal produces toxic waste. Processed coal waste, or slurry, is stored in more than 500 impoundments across the nation. The most spectacular failures of slurry impoundments occurred at Buffalo Creek, West Virginia, in 1972, when 125 people were killed and 1,000 injured during a flash flood caused by a failed slurry dam; in 2000, near Inez, Kentucky, when an impoundment managed by Massey Energy subsidiary Martin County Coal collapsed and spilled 300 million gallons of coal sludge into neighboring Kentucky towns and two creeks leading into the Big Sandy River; and in 2008, when a Tennessee Valley Authority retaining wall collapsed, spilling 1.1 billion gallons of sludge into a nearby community and the Emory River. See U.S. Environmental Protection Agency, "EPA Releases Survey Results on Coal Ash Impoundments," September 8, 2009, www.epa.govaging/press/epanews/2009/2009_0908_2.htm; "Marshall University's Virtual Museum: Buffalo Creek Flood, 1972," 2008, www.marshall.edu/library/libassoc/buffalocreek.asp; Roger Alford, "Coal Sludge Lingers from Martin County Spill," *Lexington (Ky.) Herald-Leader*, June 17 2003, www.kentucky.com; Stephanie Smith, "Months after Ash Spill, Tennessee Town Still Choking," CNN.com, July 13, 2009, www.cnn.com/2009/HEALTH/07/13/coal.ash.illnesses/index.html.

The devastating effects of mountaintop removal on the environment and public health extend far beyond what I have discussed here. For further information on mountaintop removal and coal slurry impoundments, see Michael Shnayerson, *Coal River* (New York: Farrar, Straus and Giroux, 2008). I also recommend David Novack's award-winning 2008 documentary film, *Burning the Future: Coal in America*.

75. Tim Flannery, *The Weathermakers: How Man Is Changing the Climate and What It Means for Life on Earth* (New York: Atlantic Monthly, 2005), 28.

76. Bleizeffer, "Wyo Offers Up $31.2 M Incentive for FutureGen."

77. Katie Benner, "Clean Coal: A Good Investment?," CNNMoney.com, October 19, 2004, http://money.cnn.com/2004/10/18/news/economy/coal.

78. Bleizeffer, "Panel Pushes CO_2 Business."

79. Bob Moen, "Jonah, Pinedale Gas Fields Yield Oil Too," *Casper Star-Tribune*, July 14, 2009.

80. Benner, "Clean Coal."

81. Bleizeffer, "Panel Pushes CO_2 Business."

82. Benner, "Clean Coal."

83. "Trading Hot Air," *Economist*, October 17, 2002, available at www.economist .com.

84. Bleizeffer, "Panel Pushes CO_2 Business"; "Trading Hot Air."

85. Bleizeffer, "Panel Pushes CO_2 Business."

86. "Trading Hot Air."

87. Flannery, *Weathermakers*, 225.

88. Wallace Stegner, *Where the Bluebird Sings to the Lemonade Springs: Living and Writing in the West* (New York: Penguin, 1992), 202.

89. Noah Brenner, "Wyoming Wildlife Faces Twin Threats," *High Country News*, January 24, 2005, www.hcn.org/issues/290/15229.

90. Ray Ring, "Gold from the Gas Fields," *High Country News*, November 28, 2005, www.hcn.org/issues/311/15938.

91. Jeff Gearino, "Industry Outlines Workforce Needs," *Casper Star-Tribune*, July 8, 2006.

92. Tom Mast, "Wyoming Leads Nation in Job Growth," *Casper Star-Tribune*, January 31, 2007; Tom Mast, "The Year of Job Loss," *Casper Star-Tribune*, December 27, 2009.

93. Thomas Power, "Oil and Gas Firms Know Better," *Casper-Star-Tribune*, January 31, 2010.

94. Bob Moen, "Wyoming Loses 800 Energy Jobs in 1 Month," *Casper Star-Tribune*, April 1, 2009.

95. Mast, "The Year of Job Loss."

96. Gina Vergel, "Experts Debate Future of Global Energy Consumption," news release, Fordham University, December 2009, www.fordham.edu /campus_resources/enewsroom/archives/archive_1724.asp.

97. U.S. Department of the Interior, Bureau of Land Management, High Plains District Office, "Powder River Basin Coal," January 20, 2010, www.blm .gov/wy/st/en/programs/energy/Coal_Resources/PRB_Coal.html.

98. My phone calls and e-mail messages to the Montana's managing office were not returned.

99. Kittredge, *Who Owns the West?*, 92–93.

100. Robert B. Westbrook, "In the Mirror of the Enemy: Japanese Political Culture and the Peculiarities of American Patriotism in World War II," in *Bonds of Affection: Americans Define Their Pariotism*, ed. John Bodnar (Princeton: Princeton University Press, 1996), 211.

101. National Archives and Records Administration, "Four Freedoms," *Powers of Persuasion: Poster Art from World War II*, www.archives.gov/exhibits/powers _of_persuasion/four_freedoms/four_freedoms.html.

102. Westbrook, "In the Mirror of the Enemy," 211.

103. Tom Fenton, "High-Stakes Showdown in Fallujah," CBS News, November 8, 2004, www.cbsnews.com/stories/2004/10/18/opinion/fenton/main649837 .shtml; Tony Karon, "Why Iran Has the Upper Hand in the Nuclear Showdown," *Time*, September 7, 2006, available at www.time.com.

104. Maurizio Viroli, *For Love of Country: An Essay on Patriotism and Nationalism* (New York: Oxford, 1995), 1.

105. According to the DOE, the "CAR" in FreedomCAR stands for "cooperative automotive research." As of February 2009, the partners of the FreedomCAR & Fuel Partnership included the DOE, BP America, Chevron Corporation, ConocoPhilips, Exxon Mobil Corporation, Shell Hydrogen LLC, the U.S. Council for Automotive Research, Chrysler LLC, Ford Motor Company, General Motors Corporation, DTE Energy, and California Edison. See U.S. Department of Energy, FreedomCAR and Fuel Partnership, "Addendum to the FreedomCAR and Fuel Partnership Plan to Integrate Electric Utility Industry Representatives," February 2009, www1.eere.energy.gov /vehiclesandfuels/pdfs/program/fc_fuel_addendum_2-09.pdf; see also U.S. Department of Energy, *FreedomCAR and Fuel Partnership, Partnership Plan,* March 2006, www1.eere.energy.gov/vehiclesandfuels/pdfs/program/fc_fuel _partnership_plan.pdf.

106. U.S. Department of Energy, FreedomCAR and Fuel Partnership, "Addendum."

107. U.S. Department of Energy, FreedomCAR and Fuel Partnership, Partnership Plan.

108. DaimlerChrysler, "2007 Jeep Patriot," *Jeep,* 2006, www.jeep.com (no longer available).

109. DaimlerChrysler, "Press Release: All-New 2007 Jeep Patriot Delivers Traditional Jeep Styling, Best-in-Class Capability," April 12, 2006, www .jeep.com (no longer available).

110. Richard Slotkin, *The Fatal Environment: The Myth of the Frontier in the Age of Industrialization, 1800–1890* (New York: HarperPerennial, 1994), 531.

111. Julia B. Corbett, "A Faint Green Sell: Advertising and the Natural World," in *Enviropop: Studies in Environmental Rhetoric and Popular Culture,* ed. Mark Meister and Phyllis M. Japp (Westport, Conn.: Praeger, 2002), 149.

112. Hansen, *Lost Promise of Patriotism,* 10; Bodnar, "Attractions of Patriotism," 11, 15, 16.

113. Terry Tempest Williams, *The Open Space of Democracy* (Great Barrington, Mass.: Orion Society, 2004), 86.

114. Spencer Swartz and Shai Oster, "China Tops U.S. in Energy Use," *Wall Street Journal,* July 18, 2010, available at http://online.wsj.com; Renee Montagne, "Bigger Houses Pull More Electricity for Cooling," *Morning Edition,* National Public Radio, August 2, 2006.

115. Swartz and Oster, "China Tops U.S. in Energy Use."

116. U.S. Department of Energy, Energy Information Administration, "Table 2: Natural Gas Consumption in the United States, 2004–2009," *Natural Gas Monthly,* February 2010, www.eia.doe.gov/pub/oil_gas/natural_gas/data _publications/natural_gas_monthly/current/pdf/ngm_all.pdf; U.S. Department of Energy, Energy Information Administration, "Table 6. U.S. Coal Supply, Consumption, and Inventories," March 9, 2010, www.eia.doe.gov /emeu/steo/pub/6tab.pdf; U.S. Department of Energy, U.S. Energy Information Administration, Independent Statistics and Analysis, "U.S. Crude

Oil and Liquid Fuels," *Short-Term Energy Outlook*, March 9, 2010, www.eia. doe.gov/steo#US_Crude_Oil_And_Liquid_Fuels. High as these numbers may seem, the data for oil and coal reflect dips in consumption following the sliding economy. In December 2009, the Department of Energy projected that overall fuel consumption would rise, following the trajectory of the economy. See U.S. Department of Energy, Energy Information Administration, Independent Analysis and Statistics, "Annual Energy Outlook Early Release Overview," December 14, 2009, www.eia.doe.gov/oiaf/aeo /overview.html.

117. Vergel, "Experts Debate Future of Global Energy Consumption." In discussing the International Energy Agency's *World Energy Outlook* for 2009, Ambassador Richard H. Jones outlined two consumption "scenarios": one in which nations continue on their current path, and another in which wealthier nations implement policy changes to curb global warming. Ambassador Jones said that the latter platform could reduce the number of people without access to electricity from 1.5 billion today to 1.3 billion by 2030.

118. Slotkin, *Fatal Environment*, 40.

119. Steve Inskeep, "Exploring for Oil in the 'Great Frontier,'" NPR, September 6, 2007, www.npr.org/templates/story/story.php?storyId=14194568.

120. True North Energy Corp, www.tnecorp.com.

121. Kenneth R. Weiss, "Polar Bear Listed as Endangered," *Los Angeles Times*, May 15, 2008, http://articles.latimes.com/2008/may/15/local/me-polar15.

122. Kittredge, *Who Owns the West?*, 74.

123. John Carey, "Business on a Warmer Planet," *BusinessWeek*, July 17, 2006, 28.

124. John D. Sutter, "Climate Change Threatens Life in Shishmaref, Alaska," *CNN Tech*, December 3, 2009, www.cnn.com.

125. Linda Hogan, *Dwellings: A Spiritual History of the Living World* (New York: Norton, 1995), 89.

126. "BLM Approves Pipeline Work," *Casper Star-Tribune*, July 13, 2006.

127. Susan Voyles, "Groups Push Renewable Energy for Power Line Project," *Reno Gazette-Journal*, May 23, 2005, www.rgj.com.

128. Kinder Morgan, "Rockies Express Pipeline," www.kindermorgan.com /business/gas_pipelines/rockies_express; Diane Wetzel, "$6.8 Billion Pipeline Project Completed," *North Platte (Neb.) Telegraph*, November 25, 2009, available at www.kindermorgan.com/business/gas_pipelines/rockies_express /rex _docs_news.cfm.

129. Jennifer Talhelm, "BLM Struggles With Drilling Demand," *Casper Star-Tribune*, June 29, 2006.

130. Jennifer Talhelm, "Accelerating Development," *Casper Star-Tribune*, June 7, 2006.

131. "Utah Energy Company Plans 4,000 Wells in Colorado, Wyoming," *Casper Star-Tribune*, June 29, 2006.

132. U.S. Department of Interior, Bureau of Land Management, "BLM Oil and Gas Lease Sale Nets $9.6 Million, State of Wyoming to Get Half," *Bureau of Land Management Wyoming News*, April 5 2006, www.blm.gov/wy/st/en /info/news_room/2006newsreleases/04/050gsale.htm.

133. Whitney Royster, "Wyoming Should Take Lead, Professor Says," *Casper Star-Tribune*, July 17, 2006.

134. Matt Joyce, "Wind Farm Regs Take Shape," *Casper Star-Tribune*, February 11, 2010; Dustin Bleizeffer, "Landowner, Wildlife Advocates See Opportunity in Wind Regs," *Casper Star-Tribune*, February 11, 2010.

135. University of Wyoming, "War Memorial Stadium History," www.uwyo.edu/tour/stadiumhist.htm.

136. University of Wyoming, "$5 Million Gift Names 'Jonah Field' at War Memorial Stadium," May 6, 2005, www.uwyo.edu/newsletter/2005/may/jonah.htm.

137. Ann Zwinger, *Run, River, Run: A Naturalist's Journey Down One of the Great Rivers of the American West* (1975; repr., Tucson: University of Arizona Press, 1984), 136.

138. Lewis L. Gould, *Lady Bird Johnson: Our Environmental First Lady* (Lawrence: University Press of Kansas, 1999), 39.

139. Anderson, *Thomas Moran*, 56.

140. Rick Van Noy, *Surveying the Interior: Literary Cartographers and the Sense of Place* (Reno: University of Nevada Press, 2003), 101–2.

141. Anderson, *Thomas Moran*, 56.

142. Eric Bontrager, "BLM Authorizes Grand Canyon Exploration," *New York Times*, May 6, 2009. See also U.S. Geological Survey, "USGS Report Details Uranium Resources and Potential Effects of Uranium Mining Near Grand Canyon," February 18, 2010, www.usgs.gov/newsroom/article.asp?ID=2406; and Andrea E. Alpine, ed., "Hydrological, Geological, and Biological Characterization of Breccia Pipe Uranium Deposits in Northern Arizona," U.S. Geological Survey Scientific Investigations Report 2010-5020 (February 17, 2010), http://pubs.usgs.gov/sir/2010/5025/.

143. W. J. Nuttall, *The Nuclear Renaissance: Technologies and Policies for the Future of Nuclear Power* (Philadelphia: Institute of Physics Publishing, 2005), 2. The earliest printed reference I am able to locate for the term "nuclear renaissance" is dated 1986. Hans Blix, then head of the International Atomic Energy Agency, said to an international press audience, "Chernobyl has put an end, for the time being, to this nuclear renaissance." See D. Costello, "Expert Raps Soviets on Tardy Disaster Warning," *(Brisbane, Australia) Courier-Mail*, May 15, 1986, available through LexisNexis Academic.

144. U.S. Department of Energy, "Nuclear," www.energy.gov/energysources/nuclear.htm; Barack Obama, "Remarks by the President on Energy in Lanham, Maryland," February 16, 2010, available at www.whitehouse.gov/briefing-room/speeches-and-remarks. President George W. Bush's Advanced Energy Initiative included similar language describing nuclear power.

145. Rena Delbridge, "Uranium Companies Plan Converse County Operations," *Casper Star-Tribune*, July 10, 2006.

146. Mary Oliver, *American Primitive* (Boston: Little, Brown, 1983), 77.

147. "Native American Tribes of Wyoming," *Wyoming Indian Tribes*, http://www.native-languages.org/wyoming.htm.

148. Hogan, *Dwellings*, 94.

149. Kittredge, *Who Owns the West?*, 142.
150. Peter Odell, "Conventional Wisdom Challenged," *Energy & Environment* 18.2 (March 2007): 289.
151. U.S. Environmental Protection Agency, "Clean Energy: Natural Gas," December 28, 2007, www.epa.gov/cleanenergy/energy-and-you/affect /natural-gas.html.
152. Vergel, "Experts Debate Future of Global Energy Consumption."
153. Scott Pelley, "Rewriting the Science," *Sixty Minutes*, CBS, March 19, 2006.
154. Mark Williams, "Gas Could Be the Cavalry in Global Warming Fight," Associated Press, December 20, 2009, available at http://abcnews .go.com/Business/wireStory?id=9386336.
155. Daniel Yergin, "Ensuring Energy Security," *Foreign Affairs*, March/April 2006.
156. International Energy Agency, "World Energy Outlook Press Release," November 10, 2009, www.iea.org/press/pressdetail.asp?PRESS_REL_ID =294.
157. Daniel Yergin and Robert Ineson, "America's Natural Gas Revolution," *Wall Street Journal*, November 2, 2009, available at http://online.wsj.com.
158. Clifford Krauss, "New Way to Tap Gas May Expand Global Supplies," *New York Times*, October 9, 2009; Vergel, "Experts Debate Future of Global Energy Consumption."
159. Yergen and Ineson, "America's Natural Gas Revolution."
160. Frank Luntz, *Words That Work: It's Not What You Say, It's What People Hear* (New York: Hyperion, 2007), 219.
161. Mosco, *Digital Sublime*, 117, 14.
162. U.S. Department of Energy, "DOE Awards $377 Million in Funding for 46 Energy Frontier Research Centers," August 6, 2009, www.energy.gov/7768 .htm.
163. Eric Schlosser, *Fast Food Nation: The Dark Side of the All-American Meal* (New York: Perennial, 2002), 65.
164. James Galvin, *Resurrection Update: Collected Poems, 1975–1997* (Port Townsend, Wash.: Copper Canyon, 1997), 243.
165. Stegner, *Where the Bluebird Sings*, 199–200, 205, 204.

Conclusion: Green Patriotism

1. Amy DeRogatis, *Moral Geography: Maps, Missionaries, and the American Frontier* (New York: Columbia University Press, 2003), 182.
2. Angela Miller, *The Empire of the Eye: Landscape Representation and American Cultural Politics, 1825–1875* (Ithaca, N.Y.: Cornell University Press, 1993), 3, 7–8.
3. William H. Goetzmann and William N. Goetzmann, *The West of the Imagination* (New York: Norton, 1986), 182.
4. Robert N. Bellah et al., *Habits of the Heart: Individualism and Commitment in American Life* (Berkeley: University of California Press, 1985), 335.

5. Kathleen Dean Moore, *The Pine Island Paradox: Making Connections in a Disconnected World* (Minneapolis: Milkweed Editions, 2004), 65.

6. David W. Orr, *The Last Refuge: Patriotism, Politics, and the Environment in an Age of Terror* (Washington, D.C.: Island Press, 2004) 79.

7. "Marine Corps 'The Climb,'" www.youtube.com/watch?v=Wi8RTlFxcUI.

8. John Bodnar, "The Attractions of Patriotism," in *Bonds of Affection: Americans Define Their Pariotism*, ed. Bodnar (Princeton: Princeton University Press, 1996), 11, 16.

9. Julia B. Corbett, "A Faint Green Sell: Advertising and the Natural World," in *Enviropop: Studies in Environmental Rhetoric and Popular Culture*, ed. Mark Meister and Phyllis M. Japp (Westport, Conn.: Praeger, 2002), 150.

10. Char Miller, *Gifford Pinchot and the Making of Modern Environmentalism* (Washington, D.C.: Island Press, 2001), 359; U.S. Department of Energy, "FreedomCAR and Vehicle Technologies Program: Deployment," *Energy Efficiency and Renewable Energy*, February 7, 2006, www1.eere.energy.gov /vehiclesandfuels/deployment/printable_versions/index.html.

11. Jonathan Foreman, *The Pocket Book of Patriotism* (New York: Sterling, 2005), 6.

12. William Kittredge, *Who Owns the West?* (San Francisco: Mercury House, 1996), 70.

13. Philip J. Deloria, *Indians in Unexpected Places* (Lawrence: University Press of Kansas, 2004), 7.

14. Richard Slotkin, *Gunfighter Nation: The Myth of the Frontier in the Twentieth Century* (Norman: University of Oklahoma Press, 1998), 3, 494–96.

15. William Kittredge, *Taking Care: Thoughts on Storytelling and Belief* (Minneapolis: Milkweed Editions, 1999), 77.

16. William Kittredge, *Owning It All* (St. Paul: Graywolf Press, 1987), 61.

17. Barry Lopez, Richard Nelson, and Terry Tempest Williams, *Patriotism and the American Land* (Great Barrington, Mass.: Orion Society, 2002), 13.

18. Robert F. Kennedy Jr., quoted in Oprah Winfrey, "Interview: Oprah Talks to Bobby Kennedy, Jr.," *O*, February 2007, 232.

19. Lopez, Nelson, and Williams, *Patriotism and the American Land*, 13.

20. Orr, *The Last Refuge*, 6.

21. Barbara Kingsolver, *Small Wonder: Essays* (New York: Perennial, 2002), 29.

22. Orr, *Last Refuge*, 13.

23. Chester Arnold, telephone conversation with the author, July 7, 2004. Except where noted, Arnold's comments come from this conversation.

24. Chester Arnold, e-mail to the author, June 3, 2004.

25. Christian L. Frock, "Chester Arnold Biography Presented by Catherine Clark Gallery," May 10, 2007, www.cclarkgallery.com/dynamic/artist_bio .asp?ArtistID=7.

26. Chester Arnold, quoted ibid.

27. Miller, *Empire of the Eye*, 47–48.

28. Kennecott Utah Copper, "The Pit Grows Deeper," July 23, 2004, www .kennecott.com (no longer available).

29. Ivan Weber, "Kennecott's Bad Groundwater Destined for the Great Salt Lake?," *Utah Sierran*, Fall 2000, 4.

30. Utah Department of Environmental Quality, "State of Utah Natural Resource Damage Trustee Southwest Jordan Valley Groundwater Cleanup," *Report to the Public,* June 2004, www.deq.utah.gov/Issues/nrd/docs/fdocs/Finaldraft061404drn.pdf.

31. U.S. Department of Justice, "United States and Montana Reach Agreement with Mining Companies to Clean Up Berkeley Pit," March 25, 2002, www.justice.gov/opa/pr/2002/March/02_enrd_180.htm.

32. Christopher Thorne, "Berkeley Pit Cleanup Pact Reached," *Missoulian,* March 27, 2002, available at http://missoulian.com.

33. Brian Shovers, "Remaking the Wide-Open Town: Butte, Montana, at the End of the Twentieth Century," in *Western Technological Landscapes,* ed. Stephen Tchudi (Reno: Nevada Humanities Committee / Halcyon Imprints, 1998), 153.

34. Kittredge, *Taking Care,* 76.

35. John Tirman, "The Future of the American Frontier," *American Scholar* 78.1 (Winter 2009): 30–40.

36. Maurizio Viroli, *For Love of Country: An Essay on Patriotism and Nationalism* (Oxford: Clarendon Press, 1995), 2.

37. Kittredge, *Owning It All,* 62.

38. Tirman, "The Future of the American Frontier."

39. Scott Slovic, *Seeking Awareness in American Nature Writing: Henry Thoreau, Annie Dillard, Edward Abbey, Wendell Berry, and Barry Lopez* (Salt Lake City: University of Utah Press, 1992), 169; Cheryll Glotfelty, "Literary Place Bashing, Test Site Nevada," in *Beyond Nature Writing: Expanding the Boundaries of Ecocriticism,* ed. Karla Armbruster and Kathleen Wallace (Charlottesville: University of Virginia Press, 2001), 243.

40. William Cronon, "The Trouble with Wilderness; or, Getting Back to the Wrong Nature," in *Uncommon Ground: Toward Reinventing Nature,* ed. Cronon (New York: Norton, 1995), 88.

INDEX